Small Scale Deformation Using Advanced Nanoindentation Techniques

Small Scale Deformation Using Advanced Nanoindentation Techniques

Special Issue Editors

Ting Tsui
Alex A. Volinsky

MDPI • Basel • Beijing • Wuhan • Barcelona • Belgrade

MDPI

Special Issue Editors

Ting Tsui
University of Waterloo
Canada

Alex A. Volinsky
University of South Florida
USA

Editorial Office
MDPI
St. Alban-Anlage 66
4052 Basel, Switzerland

This is a reprint of articles from the Special Issue published online in the open access journal *Micromachines* (ISSN 2072-666X) from 2018 to 2019 (available at: https://www.mdpi.com/journal/micromachines/special_issues/Small_Scale_Deformation_using_Advanced_Nanoindentation_Techniques)

For citation purposes, cite each article independently as indicated on the article page online and as indicated below:

LastName, A.A.; LastName, B.B.; LastName, C.C. Article Title. *Journal Name* **Year**, *Article Number*, Page Range.

ISBN 978-3-03897-966-1 (Pbk)
ISBN 978-3-03897-967-8 (PDF)

Contents

About the Special Issue Editors

Ting Tsui has over twenty-five years of experience in the integrated circuit (IC) fabrication industry and in the academic environment. At Texas Instruments Inc. (Dallas, Texas), Prof. Tsui developed Plasma Enhanced Chemical Vapor Deposited (PECVD) porous low-k and ultra-low-k thin films to be used for interlayer dielectric and dielectric barriers for 90 nm, 65 nm, and 45 nm technology node. He also developed plasma-based processes to enhance electrical and chip reliability performance. In addition to the unit process module development, Professor Tsui has also demonstrated his expertise in wafer-level integration, yield ramp, TDDB reliability, and in-fab and in-field mechanical reliability. In 2007, Professor Tsui began his role at the University of Waterloo (Ontario, Canada) where he has conducted research in the areas of porous ultra-low-dielectric constant materials, nano-mechanics, electron beam lithography, and nanoindentation.

Alex A. Volinsky is an Associate Professor at the University of South Florida, Mechanical Engineering Department. He obtained his Ph.D. degree from the University of Minnesota, Department of Chemical Engineering and Materials Science in 2000. Dr. Volinsky held an Engineering Materials Senior Staff Member potion at Motorola's Process and Materials Characterization Lab prior to joining USF. Professor Volinsky's current research interests are: Thin films processing, mechanical properties and characterization, adhesion and fracture of thin films, microelectronics and MEMS reliability, environmental degradation of materials. Dr. Volinsky authored over 300 scientific papers, one of which was determined to be the most cited paper in the field of Materials Science by ISI® Essential Science Indicators: A.A. Volinsky, N.R. Moody, W.W. Gerberich, Acta Mater. Vol. 50/3, pp. 441–466, 2002. He organized several conferences and symposia. Professor Volisnky's research was recognized by national and international awards.

micromachines

MDPI

Editorial

Editorial for the Special Issue on Small-Scale Deformation using Advanced Nanoindentation Techniques

Ting Tsui [1,*] and Alex A. Volinsky [2,*]

1   Department of Chemical Engineering, University of Waterloo, Waterloo, ON N2L 3G1, Canada
2   Department of Mechanical Engineering, University of South Florida, 4202 E Fowler Ave. ENB 118 Tampa, FL 33620, USA
*   Correspondence: tttsui@uwaterloo.ca (T.T.); volinsky@usf.edu (A.A.V.)

Received: 14 April 2019; Accepted: 16 April 2019; Published: 22 April 2019

check for updates

Nanoindentation techniques have been used to reliably characterize mechanical properties at small scales for the past 30 years. Recent developments of these depth-sensing instruments have led to breakthroughs in fracture mechanics, time-dependent deformations, size-dependent plasticity, and viscoelastic behavior of biological materials. This special issue contains 11 papers covering a diverse field of materials deformation behavior. Müller et al. [1] developed a new nanoindentation method to evaluate the influence of hydrogen on the plastic deformation of nickel. Effects of radiation on ferritic-martensitic steels were studied by Roldán et al. [2]. The applications of the depth-sensing indentation method in the mechanical reliability of microelectronic packaging products, such as through-silicon via (TSV) structures and lead-free solder, were performed by Wu et al. [3] and Long et al. [4], respectively. Gan et al. [5] and Chiu et al. [6] investigated the fracture behavior of cementitious cantilever beam and InP single crystals. Studies of nanometer scale deformation of metallic glass materials (Zr-Cu-Ni-Al and La-Co-Al alloys) [7] and Bi_2Se_3 thin films [8] were also part of the collected manuscripts. The mechanical deformation of mammalian cells and other biological materials [9,10] were also discussed in this focus issue. Influence of surface pit on the nanoindentation was studied by Zhang et al. [11]. The editors would like to thank these authors for their contributions to this focus issue.

References

1.   Müller, C.; Zamanzade, M. The Impact of Hydrogen on Mechanical Properties; A New In Situ Nanoindentation Testing Method. *Micromachines* **2019**, *10*, 114. [CrossRef] [PubMed]
2.   Roldán, M.; Fernández, P.; Rams, J.; Sánchez, F.J.; Adrián, G.-H. Nanoindentation and TEM to Study the Cavity Fate after Post-Irradiation Annealing of He Implanted. *Micromachines* **2018**, *9*, 633. [CrossRef] [PubMed]
3.   Wu, C.; Wei, C.; Li, Y. In Situ Mechanical Characterization of the Mixed-Mode Fracture Strength of the Cu/Si Interface for TSV Structures. *Micromachines* **2019**, *10*, 86. [CrossRef] [PubMed]
4.   Long, X.; Zhang, X.; Tang, W.; Wang, S.; Feng, Y.; Chang, C. Calibration of a Constitutive Model from Tension and Nanoindentation for Lead-Free Solder. *Micromachines* **2018**, *9*, 608. [CrossRef] [PubMed]
5.   Gan, Y.; Zhang, H.; Šavija, B.; Schlangen, E. Static and Fatigue Tests on Cementitious Cantilever Beams Using Nanoindenter. *Micromachines* **2018**, *9*, 630. [CrossRef] [PubMed]
6.   Chiu, Y.; Jian, S.; Liu, T.; Le, P.H.; Juang, J. Localized Deformation and Fracture Behaviors in InP Single Crystals by Indentation. *Micromachines* **2018**, *9*, 611. [CrossRef] [PubMed]
7.   Ma, Y.; Song, Y.; Huang, X.; Chen, Z.; Zhang, T. Testing Effects on Shear Transformation Zone Size of Metallic Glassy Films Under Nanoindentation. *Micromachines* **2018**, *9*, 636. [CrossRef] [PubMed]

8. Lai, H.; Jian, S.; Thi, L.; Tuyen, C.; Le, P.H.; Luo, C. Nanoindentation of Bi_2Se_3 Thin Films. *Micromachines* **2018**, *9*, 518. [CrossRef] [PubMed]

9. Qian, L.; Zhao, H. Nanoindentation of Soft Biological Materials. *Micromachines* **2018**, *9*, 654. [CrossRef] [PubMed]

10. Moussa, H.; Logan, M.; Wong, K.; Rao, Z.; Aucoin, M.; Tsui, T. Nanoscale-Textured Tantalum Surfaces for Mammalian Cell Alignment. *Micromachines* **2018**, *9*, 464. [CrossRef] [PubMed]

11. Zhang, Z.; Ni, Y.; Zhang, J.; Wang, C.; Ren, X. Multiscale analysis of size effect of surface pit defect in nanoindentation. *Micromachines* **2018**, *9*, 298. [CrossRef] [PubMed]

micromachines

MDPI

Communication

The Impact of Hydrogen on Mechanical Properties; A New In Situ Nanoindentation Testing Method

Christian Müller [1], Mohammad Zamanzade [1,2,*] and Christian Motz [1]

[1] Department of Materials Science and Methods, Saarland University, Bldg. D2.2, 66123 Saarbrücken, Germany; c.mueller@matsci.uni-saarland.de (C.M.); motz@matsci.uni-sb.de (C.M.)
[2] Mines Saint-Etienne, Univ Lyon, CNRS, UMR 5307LGF, Centre SMS, F-42023 Saint-Etienne, France
* Correspondence: mohammad.zamanzade@emse.fr; Tel.: +33-0-477420048

Received: 13 January 2019; Accepted: 6 February 2019; Published: 11 February 2019

check for
updates

Abstract: We have designed a new method for electrochemical hydrogen charging which allows us to charge very thin coarse-grained specimens from the bottom and perform nanomechanical testing on the top. As the average grain diameter is larger than the thickness of the sample, this setup allows us to efficiently evaluate the mechanical properties of multiple single crystals with similar electrochemical conditions. Another important advantage is that the top surface is not affected by corrosion by the electrolyte. The nanoindentation results show that hydrogen reduces the activation energy for homogenous dislocation nucleation by approximately 15–20% in a (001) grain. The elastic modulus also was observed to be reduced by the same amount. The hardness increased by approximately 4%, as determined by load-displacement curves and residual imprint analysis.

Keywords: nickel; nanoindentation; hardness; brittleness and ductility; hydrogen embrittlement

1. Introduction

Conventional mechanical testing methods have been used for quantitative studies of the influence of hydrogen on mechanical properties, e.g., yield stress, ultimate tensile stress and fracture strain [1]. However, these techniques are not very successful in obtaining mechanistic information because they simultaneously probe a large volume of the material and only provide an averaged result as if the volume were homogeneous. In fact, macroscopic samples contain inhomogeneities such as vacancies, dislocations and grain boundaries, which are known to play an important role in hydrogen embrittlement [2–4]. In contrast to macroscopic experiments, local testing methods enable us to decrease the probed volume of material, perform measurements for a quasi-homogeneous volume of material and hence decrease the possible sources of scattering in the results [5–7]. Additionally, as a result of the small probed volume, the hydrogen concentration can be assumed to be constant. In the past, we have used various local, in situ experimental techniques, such as electrochemical nanoindentation, in situ compression and bending of micro-pillars, to study the contribution of solute hydrogen on elastic properties, dislocation nucleation and hardness of alloys. These techniques enabled us to achieve an understanding of the mechanisms of hydrogen embrittlement (HE) for a material in a certain medium. Furthermore, we were able to rank the sensitivity of different alloys to hydrogen embrittlement in a specific medium irrespective of the impacts of grain boundaries, phase boundaries, pores, etc. [5–7].

However, there is still room for improvement to make electrochemical setups easier and experiments more reproducible. Our previous in situ nanoindentation approach comprised a layer of electrolyte on the specimen surface, which was penetrated by the tip during indentation. This had several disadvantages: (i) capillary forces acting on the tip; (ii) an inability to use the optical microscope of the machine for positioning; (iii) either an electrolyte flow causing vibrations or a static electrolyte film, in which no chemical reaction products are washed away from the surface; (iv) corrosion of the tested surface and, accordingly, (v) limited time for hydrogen charging as well as mechanical testing. In this paper, a new testing setup for studying the impact of hydrogen on mechanical properties is introduced. With this method, hydrogen is provided at the bottom surface of a thin sample while nanoindentation is performed on the top surface. This method avoids most of the problems named above and also allows the analysis of previously unobtainable properties with respect to diffusion.

2. Experiments

Pure (99.9%) polycrystalline, square-shaped nickel specimens were milled from the back, heat-treated at 1240 °C for grain growth, ground in multiple steps, and finally electropolished with a solution of 1M sulfuric acid solved in methanol. These steps minimized the possible amount of residual stress and plastic deformation in the material, especially near the surface, where no sources for inhomogeneous dislocation nucleation were desired. The resulting specimen geometry had a thickness of approximately 200 μm in the region of interest (marked as the area with superimposed electron backscatter diffraction (EBSD) map in Figure 1), which was smaller than the average grain diameter (~500 μm). The sample was then glued to the dedicated cell made from polyvinyl chloride (PVC). The holes in the base plate of the cell allowed us to securely install it in the indenter, ensuring it was positioned identically before and after hydrogen permeation. The detailed specimen preparation and configuration of the electrochemical setup have been published elsewhere [8].

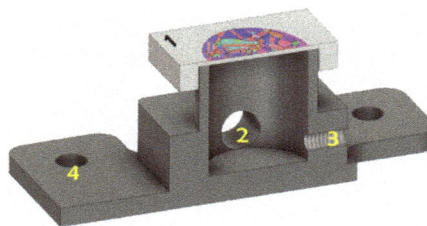

Figure 1. Cross section of specimen and electrochemical cell. (1) Nickel specimen, (2) electrolyte in-/output, (3) screw as counter electrode (4) holes for fixation in the nanoindenter.

The first electrochemical hydrogen charging step was carried out ex situ for two days with a constant current density of 14.5 A/m^2, applied by an IVIUM CompactStat (Ivium Technologies B.V., Eindhoven, The Netherlands). A 0.25 molar H$_2$SO$_4$ solution was used as the electrolyte, containing 5 g/L potassium iodide (KI) as hydrogen recombination poison. To remove hydrogen bubbles at the charging surface, the sample was inverted and the solution was steadily pumped in and out. The risk of outgassing was accounted for by charging in situ during measurements. For this purpose, a lower current density of 0.25 A/m^2 was applied and the flow rate of the pumped solution was decreased to avoid vibrations. Furthermore, a borate buffer solution (mixed using 1.24 g/L H$_3$BO$_3$ and 1.91 g/L Na$_2$B$_4$O$_7$·10H$_2$O, also supplemented with 5 g/L KI) was used instead of sulfuric acid. However, the outgassing of hydrogen may also be considerably reduced by the existence of a homogeneous passive layer [8,9], which was measured to be approximately 10 nm, using a JEOL JEM-ARM200F transmission electron microscope (TEM).

As a proof of concept, a (001)-oriented grain was tested before and after hydrogen charging. Nanoindentation was performed with a Hysitron Triboindenter (Bruker Corporation, Billerica, MA, USA), equipped with Performech controller and a diamond Berkovich tip with a tip radius of approximately 400 nm. The applied indentation parameters are summarized in Table 1, where "fast" and "slow" corresponds to the loading rate.

Table 1. Nanoindentation parameters.

Measured Parameter	Max. Load	Load or Strain Rate	Type of Test	# of Indents
Pop-in slow	700 μN	50 μN·s^{-1}	Quasi static	60
Pop-in fast	700 μN	5000 μN·s^{-1}	Quasi static	60
Hardness fast	10 mN	0.5 s^{-1}	Quasi static	10
Hardness slow	10 mN	0.05 s^{-1}	Dynamic NanoDMAIII	10

3. Results and Discussions

We observed an elastic deformation of the thin membrane during nanoindentation despite the low applied forces. Quantitative analysis of the results was performed after precise evaluation of total system compliance, which can be interpreted as a series connection of three springs: (i) compliance of the machine frame, (ii) PVC cell and the glues used to install the sample, and (iii) the deflection of the thin nickel membrane. The first calibration was a standard calibration involving the indentation of a fused quartz reference with high forces. The second spring constant was evaluated by testing a bulky nickel sample installed on the same cell and attached with the same glue. To evaluate the third spring constant or the effect of the bending of the whole membrane on the nanoindentation curves, the continuous stiffness method (Hysitron NanoDMA measurement) was used. This technique measures the elastic modulus at every indentation depth by continuously oscillating the tip. On a reference specimen, we verified that the modulus of the nickel bulk is independent of depth. Hence, the internal compliance value in all other data files were modified until their NanoDMA results also met this requirement.

Figure 2 shows load-displacement curves before and after charging. The pop-in or displacement burst phenomenon was related to the nucleation of dislocations around maximum shear stress under the tip [10]. The probability of the pop-in event did not change after hydrogen charging (in both cases more than 95%). The few curves in which no distinct pop-in could be detected are not displayed. A 15% reduction of average P_{pop-in} values was determined to be attributable to hydrogen charging as well as a slight increase in scattering. The scattering in P_{pop-in} is a common observation in nanoindentation experiments, originating from the thermal activation of homogenous dislocation nucleation [11,12].

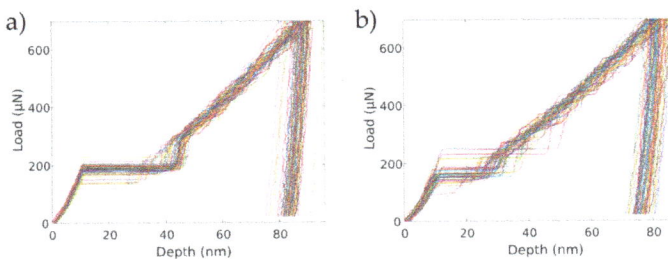

Figure 2. Load-displacement curves recorded with the fast loading rate (**a**) before charging and (**b**) after charging with hydrogen. Curves in which no pop-in could be detected were filtered.

Results indicate that the critical energy needed for dislocation nucleation was decreased, which agrees with previous studies [5,7,13]. Because of the slight reduction of the average values of pop-in load and because the population of the pop-in event did not change after charging, we can assume that the dislocation nucleation was homogeneous, and the observed behavior could be related to the debonding effect of hydrogen, known as hydrogen-enhanced decohesion (HEDE).

Another consequence of decohesion is the reduction of the elastic modulus, which was noticeable as a reduction of the slope in the Hertzian regimes, where the curves follow the equation [14]:

$$P = \frac{4}{3}E_r\sqrt{R} \times h^{3/2} \tag{1}$$

in which E_r is the reduced elastic modulus, R is the tip radius, P. is the applied load, and h the indentation depth. The load-displacement $(P - h)$ data before the pop-in can be fitted to this equation with a fit parameter proportional to E_r. To visualize the differences and make them independent of R, a cumulative distribution of relative elastic moduli E_r/E_0 is plotted in Figure 3a, where E_0 is the average value before hydrogen charging. The reduction of reduced elastic modulus according to this analysis is approximately $17 \pm 10\%$, from 206 ± 20 GPa to 171 ± 17 GPa. Both individual values are in a realistic range for nickel. The reliability of each data point obtained by curve fitting and the deviation between individual points account for the uncertainty of this result. A more conventional method to determine Young's modulus in nanoindentation experiments is to fit the unloading segment with the model introduced by Oliver and Pharr [15]. Using this approach for our measurements, the moduli before and after charging were both approximately 200 GPa and were well inside the statistical error interval of each other.

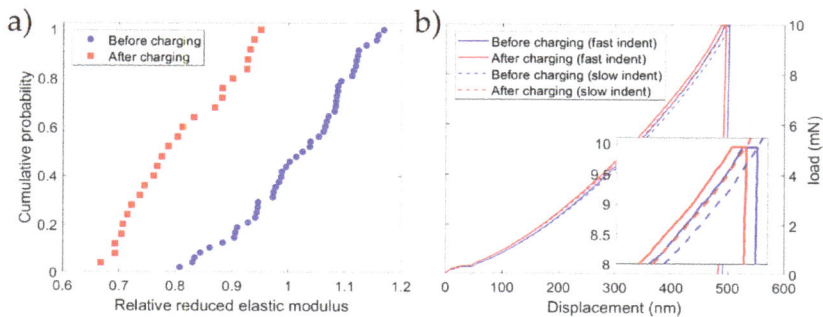

Figure 3. (a) Plot of reduced elastic modulus determined from Hertzian loading. (b) Load-displacement data of hardness measurements (averaged across 10 indents).

The occurrence of a measurable difference in Young's moduli due to hydrogen is controversial. For example, Lawrence et al. determined a comparable reduction by about 22% in nanoindentation experiments on nickel with the conventional Oliver-Pharr method [16]. In other studies, the difference was much smaller or even negligible, e.g., in molecular dynamics simulations of hydrogen in nickel [17]. Nevertheless, the analysis of elastic moduli is very much possible in our setup after the previously mentioned compliance correction.

Using the reduction of both $P_{\text{pop-in}}$ and E_r, we can also calculate the reduction of the shear stress necessary for dislocation nucleation. According to the Hertzian model, the maximum shear stress under the tip is given by:

$$\tau_{\text{max}} = 0.31 \times \left(\frac{6PE_r^2}{\pi^3 R^2} \right)^{1/3} \tag{2}$$

in which both P and E_r are reduced by 15–17%, so the stress decreases by the same amount. The calculated values of approximately 5–7 GPa are close to the expected theoretical shear strength of nickel [18], which confirms that the cause of the pop-in was indeed homogenous dislocation nucleation.

Figure 3b shows the *P-h* curves performed to study the impacts of strain rates and dissolved hydrogen on the plastic behavior of Ni. Each curve is an average of ten indents, which was calculated and plotted. Similar to other solute atoms, hydrogen can contribute to the pinning of dislocations by forming Cottrell atmospheres around dislocation cores [19,20]. The resulting reduction of dislocation mobility becomes clear if we compare the hardness before and after hydrogen charging. Measurements after hydrogen charging systematically show a decreased indentation depth at the maximum load. Increasing the loading rate also results in a smaller depth and therefore higher hardness, but this effect appears to be independent of hydrogen concentration. After measuring the projected area of the residual imprint of every indent, we determined a 4.3% increase in hardness due to hydrogen charging. Our results indicate that the application of various strain rates does not change this pining effect. Accordingly, the dislocations are constantly aged to saturation at room temperature at both tested strain rates. This means that when a dislocation escapes from a pinning point, the hydrogen diffusion is fast enough (relative to the dislocation velocity) to immediately follow and pin it again.

Another observation that reinforces the proposition of reduced dislocation mobility is a substantial reduction of pop-in width (also called excursion length) after hydrogen charging. As Figure 4 shows, a pop-in that happened at the same load P would, on average, cause a much smaller excursion length. This is attributable to the dislocations which show a drag force caused by hydrogen, decelerating the tip so that it stops earlier. Although excursion length is often assumed to solely represent the number of nucleated dislocations [21], we believe that the gliding of dislocations plays an important role in the measured pop-in width. Therefore, although hydrogen can ease the nucleation of dislocations, the reduction of the slope of the curves in Figure 4 after hydrogen charging could be related to the sessile behavior of dislocations.

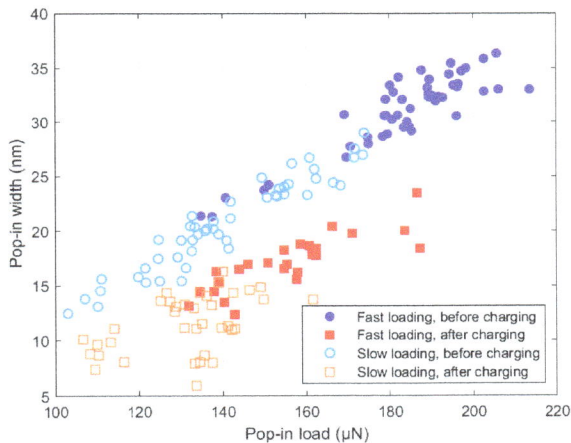

Figure 4. The impact of loading rate and hydrogen on the pop-in width and pop-in load.

4. Conclusions

The increase in hardness indicates a decreased mobility of dislocations due to the solute drag effect of hydrogen. However, hydrogen charging reduces the elastic modulus and the pop-in load and, accordingly, facilitates the formation of dislocations. The tested specimen geometry and charging conditions have proven to be successful and may be able to further analyze the influences of grain orientations and grain boundaries resulting from the testing of a thin layer charged from the back. The concept is promising for future research on diffusion coupled with changes in the mechanical properties of a variety of conductive materials.

One disadvantage introduced by the proposed setup is the consequence of the thin specimen geometry. The coarse-grained nickel layer is prone to deformation. Although its influence could be easily eliminated from nanoindentation curves, it leads to a long preparation procedure in which care must be taken in every step. Some of the electrochemical uncertainties also remain: The bottom of the sample may still be exposed to corrosion and the local hydrogen concentration at a certain point on the top can depend on inhomogeneities through which the hydrogen has to diffuse. In combination with potential discrepancies in delays between ex situ and in situ charging (and therefore outgassing), this remains a limitation on the quantitative reproducibility of specific results.

Micromachines **2019**, *10*, 114

Author Contributions: Conceptualization, M.Z.; Methodology, M.Z. and C.M. (Christian Müller); Software, C.M. (Christian Müller); Validation, M.Z., C.M. (Christian Müller) and C.M. (Christian Motz); Formal Analysis, C.M. (Christian Motz); Investigation, M.Z. and C.M. (Christian Motz); Resources, M.Z., C.M. (Christian Müller) and C.M. (Christian Motz); Data Curation, C.M. (Christian Müller); Writing-Original Draft Preparation, C.M. (Christian Müller) and M.Z.; Writing-Review & Editing, C.M. (Christian Motz); Visualization, C.M. (Christian Müller); Supervision, C.M. (Christian Motz); Project Administration, M.Z., C.M. (Christian Müller) and C.M. (Christian Motz); Funding Acquisition, M.Z.

Funding: This research was funded by the Deutsche Forschungsgemeinschaft (DFG) grant number ZA 986/1-1.

Acknowledgments: Workshop manager Stefan Schmitz and Peter Limbach, TEM operator Jörg Schmauch, proof reader Isabelle Wagner.

Conflicts of Interest: The authors declare no conflict of interest.

References

1. Gangloff, R.P.; Somerday, B.P. *Gaseous Hydrogen Embrittlement of Materials in Energy Technologies: Mechanisms, Modelling and Future Developments*; Elsevier: Amsterdam, The Netherlands, 2012; ISBN 0857095374.
2. Angelo, J.E.; Moody, N.R.; Baskes, M.I. Trapping of hydrogen to lattice defects in nickel. *Model. Simul. Mater. Sci. Eng.* **1995**, *3*, 289–307. [CrossRef]
3. Besenbacher, F.; Myers, S.M.; Nørskov, J.K. Interaction of hydrogen with defects in metals. *Nucl. Instruments Methods Phys. Res. Sect. B Beam Interact. Mater. Atoms* **1985**, *7*, 55–66. [CrossRef]
4. Myers, S.M.; Baskes, M.I.; Birnbaum, H.K.; Corbett, J.W.; DeLeo, G.G.; Estreicher, S.K.; Haller, E.E.; Jena, P.; Johnson, N.M.; Kirchheim, R. Hydrogen interactions with defects in crystalline solids. *Rev. Mod. Phys.* **1992**, *64*, 559. [CrossRef]
5. Barnoush, A.; Vehoff, H. Recent developments in the study of hydrogen embrittlement: Hydrogen effect on dislocation nucleation. *Acta Mater.* **2010**, *58*, 5274–5285. [CrossRef]
6. Zamanzade, M.; Barnoush, A. An overview of the hydrogen embrittlement of iron aluminides. *Procedia Mater. Sci.* **2014**, *3*, 2016–2023. [CrossRef]
7. Barnoush, A.; Zamanzade, M. Effect of substitutional solid solution on dislocation nucleation in Fe$_3$Al intermetallic alloys. *Philos. Mag.* **2012**, *92*, 3257–3268. [CrossRef]
8. Zamanzade, M.; Müller, C.; Velayarce, J.R.; Motz, C. Susceptibility of different crystal orientations and grain boundaries of polycrystalline Ni to hydrogen blister formation. *Int. J. Hydrogen Energy* **2019**. Accccepted for publication.
9. Condon, J.B.; Schober, T. Hydrogen bubbles in metals. *J. Nucl. Mater.* **1993**, *207*, 1–24. [CrossRef]
10. Wen, M.; Zhang, L.; An, B.; Fukuyama, S.; Yokogawa, K. Hydrogen-enhanced dislocation activity and vacancy formation during nanoindentation of nickel. *Phys. Rev. B* **2009**, *80*, 094113. [CrossRef]
11. Gao, X. Displacement burst and hydrogen effect during loading and holding in nanoindentation of an iron single crystal. *Scr. Mater.* **2005**, *53*, 1315–1320. [CrossRef]
12. Zamanzade, M.; Hasemann, G.; Motz, C.; Krüger, M.; Barnoush, A. Vacancy effects on the mechanical behavior of B2-FeAl intermetallics. *Mater. Sci. Eng. A* **2018**, *712*, 88–96. [CrossRef]
13. Zamanzade, M.; Vehoff, H.; Barnoush, A. Cr effect on hydrogen embrittlement of Fe$_3$Al-based iron aluminide intermetallics: Surface or bulk effect. *Acta Mater.* **2014**, *69*, 210–233. [CrossRef]
14. Hertz, H. On the contact of elastic solids. *J. Reine Angew Math.* **1881**, *92*, 156–171.
15. Oliver, W.C.; Pharr, G.M. An improved technique for determining hardness and elastic modulus using load and displacement sensing indentation experiments. *J. Mater. Res.* **1992**, *7*, 1564–1583. [CrossRef]
16. Lawrence, S.K.; Somerday, B.P.; Karnesky, R.A. Elastic property dependence on mobile and trapped hydrogen in Ni-201. *Jom* **2017**, *69*, 45–50. [CrossRef]
17. Wen, M.; Barnoush, A.; Yokogawa, K. Calculation of all cubic single-crystal elastic constants from single atomistic simulation: Hydrogen effect and elastic constants of nickel. *Comput. Phys. Commun.* **2011**, *182*, 1621–1625. [CrossRef]
18. Černý, M. *Theoretical Strength and Stability of Crystals from First Principles*; VUTIUM: Brno, Czech Republic, 2008.
19. Cottrell, A.H.; Jaswon, M.A. Distribution of solute atoms round a slow dislocation. *Proc. R. Soc. Lond. A.* **1949**, *199*, 104–114.

20. Delafosse, D. *Hydrogen Effects on the Plasticity of Face Centred Cubic (fcc) Crystals*; Woodhead Publishing: Sawston, Cambridge, UK, 2012; ISBN 9780857095367.
21. Bahr, D.F.; Kramer, D.E.; Gerberich, W.W. Non-linear deformation mechanisms during nanoindentation. *Acta Mater.* **1998**, *46*, 3605–3617. [CrossRef]

micromachines

MDPI

Article

In Situ Mechanical Characterization of the Mixed-Mode Fracture Strength of the Cu/Si Interface for TSV Structures

Chenglin Wu *, Congjie Wei and Yanxiao Li

Department of Civil, Architectural, and Environmental Engineering,
Missouri University of Science and Technology, Rolla, MO 65409, USA;
cw6ck@mst.edu (C.W.); yl42y@mst.edu (Y.L.)
* Correspondence: wuch@mst.edu; Tel.: +1-573-341-4465

Received: 7 December 2018; Accepted: 21 January 2019; Published: 25 January 2019

check for
updates

Abstract: In situ nanoindentation experiments have been widely adopted to characterize material behaviors of microelectronic devices. This work introduces the latest developments of nanoindentation experiments in the characterization of nonlinear material properties of 3D integrated microelectronic devices using the through-silicon via (TSV) technique. The elastic, plastic, and interfacial fracture behavior of the copper via and matrix via interface were characterized using small-scale specimens prepared with a focused ion beam (FIB) and nanoindentation experiments. A brittle interfacial fracture was found at the Cu/Si interface under mixed-mode loading with a phase angle ranging from 16.7° to 83.7°. The mixed-mode fracture strengths were extracted using the linear elastic fracture mechanics (LEFM) analysis and a fracture criterion was obtained by fitting the extracted data with the power-law function. The vectorial interfacial strength and toughness were found to be independent with the mode-mix.

Keywords: TSV; nanoindentation; FIB; micro-cantilever beam; mixed-mode; fracture

1. Introduction

Thermal mechanical reliability plays a critical role in microelectronic devices, affecting their performance and service life spans. In situ mechanical characterizations are essential to predict the thermal–mechanical behaviors of these devices. The associated techniques and approaches rapidly emerge along the technology growth in 3D integrated circuits and devices [1–7]. One of the typical approaches is nanoindentation [6,8,9], which utilizes a small-scale probe with controlled force and displacement applied directly to the substrates or micro- and nanostructures [10,11]. Utilizing various sizes and shapes of the probe, the small-scale nonlinear material behavior can be characterized. This work focuses on the latest development of the nanoindentation techniques applied to 3D integrated microelectronic devices with a through-silicon via (TSV).

As microelectronic devices become smaller and more complex, 3D integration becomes necessary for more efficient engineering and design. This integration consists of the micrometer copper vias passing through silicon die, serving as both electronical connections and mechanical supports. The copper vias are typically deposited by the electroplating approach and have complex grain structures. Under such conditions, the TSVs share different material properties, comparing to the bulk copper. Surface treatments are often conducted to the TSVs to avoid diffusion and enhance mechanical strength at Cu/Si interface. To have a comprehensive understanding of the mechanical behavior of the TSV and related interface, in situ small scale characterizations are required.

Nanoindentations have been widely adopted for in situ characterization of mechanical properties of thin-films and nanostructured materials [6–11]. The elastic and plastic properties can be readily extracted using the force–displacement responses produced by nanoindentation with various tip shapes and sizes [12–14]. In addition, miniature specimens prepared using focused ion beam (FIB) fabrication techniques can also be utilized to obtain a more systematic understanding of the deformation mechanisms at small-scales. Therefore, the combination of nanoindentation and FIB fabrication presents a unique opportunity in probing the mechanical behavior of TSV structures and interfaces in 3D integrated microelectronic devices. In this paper, a cantilever beam approach for extracting the mixed-mode interface strength is proposed. Miniature cantilever beams with various lengths were fabricated using a FIB. Both analytical and numerical models were developed to extract the mixed-mode interfacial strength at the TSV/Si interface. The extracted results were then fitted with the power-law failure criterion [15–18] producing an input for failure prediction and reliability evaluations.

2. Materials and Sample Preparation

The as-received TSV structure has periodic blind Cu arrays in a (001) Si wafer with a depth of 780 μm. The nominal via diameter and depth were 10 and 55 μm with a pitch spacing of 40 μm along the (110) direction and 50 μm along the (100) direction of the wafer, as illustrated in Figure 1. Two types of miniature specimens were prepared: The micro-pillar and cantilever beam specimens. The micro-pillar specimens were prepared by dicing and polishing the silicon wafer to have one row of the via away from the free surface by a distance of 20 μm. For each micro-pillar specimen, the top 100 nm was removed to avoid the effect of surface roughness. The silicon around the selected via was then subsequently removed, following a pattern of a concentric ring with a 3 μm thickness, as illustrated in Figure 1e. The inner ring was set at the same size as the via diameter, the outer ring was then about 16 μm in diameter. Due to the tapering effect, the top diameter of the via after the milling was about 6 μm, which formed 2 degrees of tapering angle along the via length. The micro-cantilever beam specimens were milled out of the silicon matrix near the copper via using a similar beam energy (ranging from 3–300 keV) used for the micro-pillar specimens. The side view of the prepared micro-cantilever beam is shown in Figure 1f. More details of the fabricated micro-cantilever beam are shown in Figure 2. A total of six types of micro-cantilever beam specimens were prepared with various lengths ranging from 1 to 30 μm. The width and height of the beam were set to be close to 1 μm. A specially designed square loading pad was also fabricated at the end of the beam with a size of 5.1 μm (note that the length of the loading pad was excluded from the total length to obtain the beam length). A probing crater with a diameter of 2.5 μm was carved into the loading pad to avoid the slipping of indenter tip during loading. At the Cu/Si interface, a pre-milled notch with a length of 100 nm was created, serving at the pre-crack. A total of 3 specimens were fabricated for each type of the micro-cantilever beams.

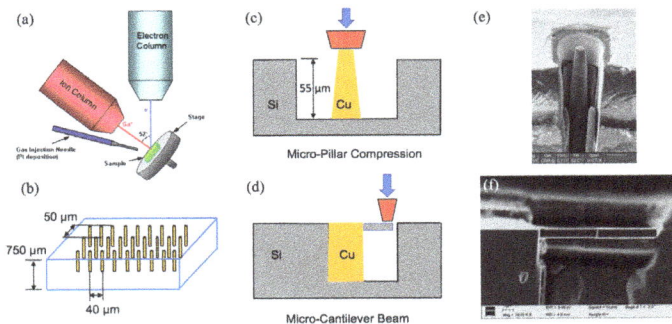

Figure 1. Through-silicon-via (TSV) specimens: (**a**) Focused ion beam scanning electron microscopy (FIB-SEM) dual beam system, (**b**) TSV in silicon substrate, schematics of (**c**) micro-pillar, (**d**) micro-cantilever experiments, SEM images of (**e**) micro-pillar adapted with permission from [8], (**f**) cantilever beam specimens.

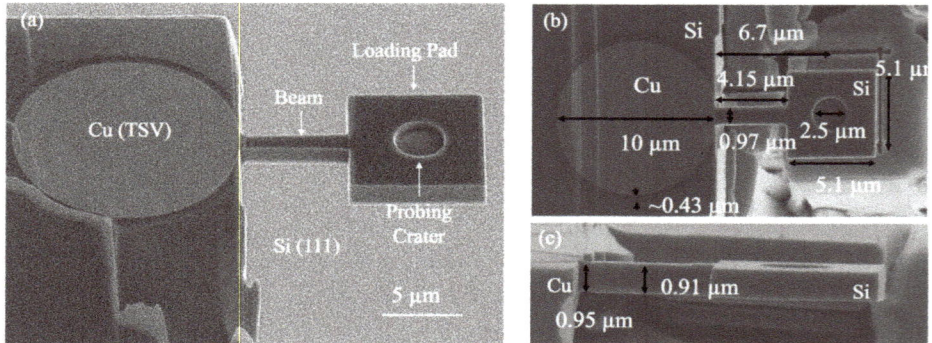

Figure 2. Micro-cantilever beam specimens: (**a**) Isometric view and structural components, (**b**) top view, and (**c**) side view with dimensional details (*L* = 4 μm).

3. Nanoindentation Experiment

The nanoindentation experiments were conducted using the Hysitron TI-95 Tribo-indenter® (Bruker Corporation, Billerica, MA, USA) on micro-cantilever specimens with a flat-punch tip having diameters of 2 μm. The micro-pillar results for analysis were obtained from our previous work [8]. The experimental details and subsequent extraction methods have been described in our previous work [8,19]. For the micro-cantilever beam experiment, the flat punch tip was placed inside the loading crater of the loading pad to apply displacement-controlled loading. A loading rate of 0.5 nm/s was applied until the contact between the cantilever beam and the sample's surface was reached. It is worth noting here that the mechanical backlash was corrected during the tip-optic calibration process. A pre-loading with a maximum load of 1 μN was applied at the end of the beam to ensure proper contact.

4. Analysis

4.1. Plastic Behavior of Cu

The force–displacement response obtained from a previous experiment [8] is shown in Figure 3a. Observing these results from the previous work, significant plastic responses were observed, as indicated by the permanent deformation after each unloading. As explained in the previous work [8], the residual deformations were also confirmed by the SEM images, shown in Figure 3a. To extract this observed elastic–plastic property, a finite element analysis was conducted, considering the tapering caused by non-uniform stress distribution [8,19]. The Ramberg–Osgood power-law relationship [20] was adopted in the numerical models to compare with experimental results. The J-2 flow theory was used to model the Cu plasticity. The 4-node quadrilateral axis-symmetrical elements in commercial finite element code ABAQUS®(Abaqus Inc., Providence, RI, USA) were used for the finite element modeling. The Ramberg–Osgood power-law relationship has been widely used for the description of plastic strain hardening of nanoindentation experiments, the stress versus plastic strain curve based on this law showed good agreement with experimental data [21–23]. In this relationship, the stress versus plastic strain response follows the description below,

$$\varepsilon_p = \frac{3}{7} \frac{\sigma_e}{E} \left(\frac{\sigma_e}{\sigma_0} \right)^{n-1} \tag{1}$$

where ε_p is the plastic strain, σ_e is the equivalent stress, σ_0 is the yield stress (which is found to be around 216 MPa), n is the Ramberg–Osgood parameter—which was found to be three from the fitting results [8,19]—and E is Young's modulus (which is found to be 110 GPa), were obtained with the

Oliver–Pharr approach, using a conical probe has a tip radius of 500 nm. This method is well applied to axis-symmetrical indenter geometries. The reduced modulus is given by:

$$E_r = \frac{\sqrt{\pi}}{2\sqrt{A(h_c)}} S \tag{2}$$

where, $S = \left(\frac{dP}{dh}\right)_{P_{max}}$ is the contact stiffness obtained from test data, $A(h_c)$ is the contact area at contact depth, h_c, given by $h_c = h_{max} - \frac{\varepsilon P_{max}}{S}$. ε equals 1 for flat-ended punch. The Young's modulus can then be obtained with:

$$E = \frac{1 - v^2}{\frac{1}{E_r} - \frac{1 - v_i^2}{E_i}} \tag{3}$$

where v and v_i are the Poisson ratio of sample and indenter, respectively.

The extracted elastic–plastic properties of the copper were used to evaluate the fracture strength at the Cu/Si interface. More analysis details are shown in Figure 4, where both the von Mises stress and equivalent plastic strains show non-uniform distributions. The non-uniform distribution was an indication of the tapering effect and further demonstrated the needs of conducting finite element analysis (FEA) to extract the plastic properties of Cu. This result also shows that the nonlinear force–displacement response has geometrical effects. Stress and plastic strain contours for FE modeling of micro-pillar compression are shown in Figure 4, where yield stress and the Ramberg–Osgood parameter are taken to be 216 MPa and 3, respectively. The slight gradient shown in the contour plots was due to the tapered cross-section of the specimen.

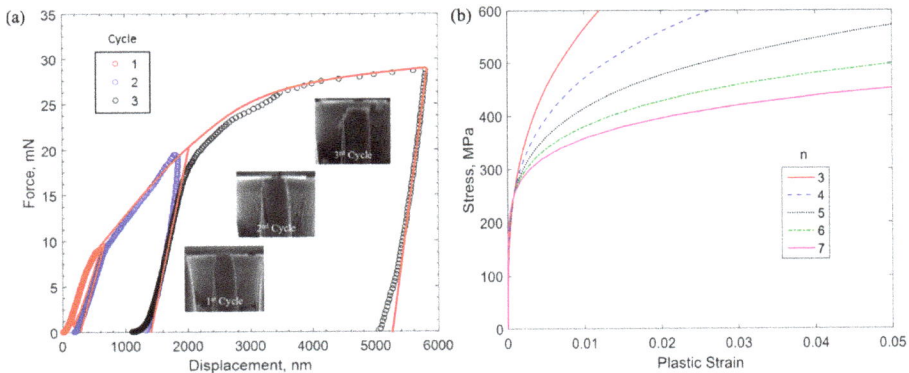

Figure 3. (**a**) Force–displacement response of micro-pillar experiment, reproduced with permission from [8], and (**b**) stress–plastic strain relationship from Ramberg-Osgood relationship.

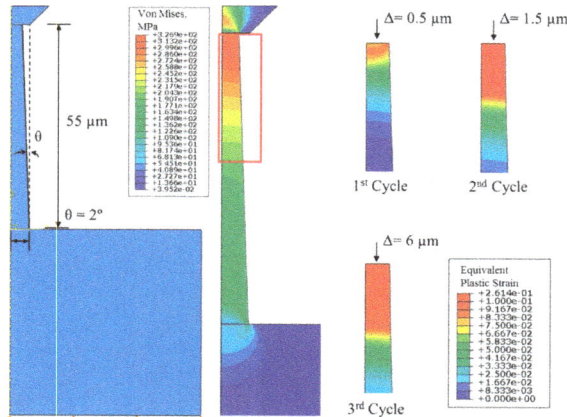

Figure 4. Finite element analysis of micro-pillar experiment, reproduced with permission from [8].

4.2. Micro-Cantilever Experiment

4.2.1. Failure Surface Characterization

The force versus displacement response for a typical micro-cantilever beam specimen is shown in Figure 5. The early contact was established as shown by the turning point between the approaching and loading response. A linear response was observed followed by a sudden failure, in terms of the drop of the force from the peak value to zero. This sudden force drops indicated a brittle Cu/Si interface. The failed surface shown in Figure 6 was characterized using SEM and energy dispersive spectroscopy (EDS) as labeled out with the red box. The elements and weight percentage results are shown in Table 1. As listed, most of the elements detected were Cu, which was followed by Si and elements in the liner materials at the TSV/Cu interface (Fe, Ta, Os). This result is similar to that of the shear failure surface from the previous work [8]. As previously concluded, the majority of the Cu signal comes from the background Cu materials in the TSV, which indicated an interfacial failure locus within the silicon matrix. The Young's modulus for Cu and Si are 110 and 165 GPa, respectively. The shear modulus used in the analysis for Cu and Si are 42.3 and 64.45 GPa, respectively, the Poisson's ratios are 0.3 and 0.28, respectively.

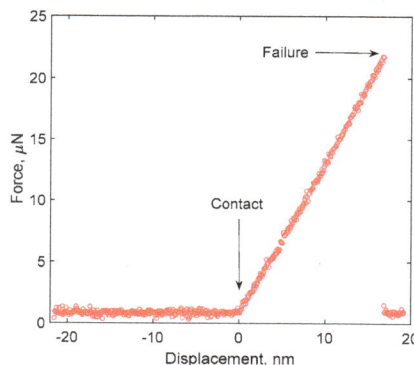

Figure 5. Typical force-displacement response of micro-cantilever beam ($L = 4\ \mu m$).

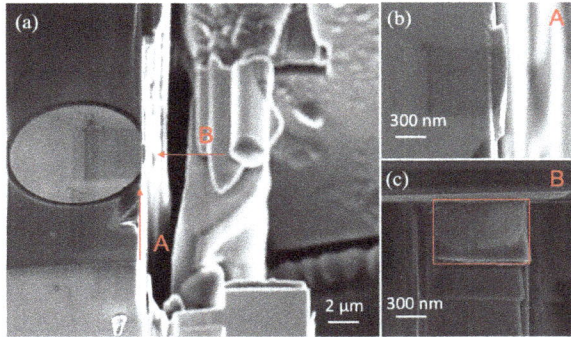

Figure 6. (a) SEM images of post-failure of cantilever beam, (b) top view details, (c) interfacial details (element analysis conducted within the red-boxed region).

Table 1. Element and weight percentage of energy dispersive spectroscopy (EDS).

Element	Weight %
Si	5.62
Fe	0.02
Cu	84.34
Ta	1.04
Os	0.01

4.2.2. Mixed-Mode Fracture

(1) Linear Elastic Fracture Mechanics (LEFM) Analysis

The stress analysis for the micro-cantilever beam experiment was conducted using both LEFM and the non-linear fracture mechanical model (NLEFM), considering the effect of Cu plasticity. Analytically, the far-field load-generated near-field stress had both normal (σ_0) and shear (τ_0) components. From the Euler beam theory and ignoring the nonlinear shear deformation caused by root rotation, these stresses can be obtained using the beam geometry and material constants of the silicon.

$$\sigma_0 = \frac{PL}{6bh^2}, \ \tau_0 = \frac{3P}{2bh} \tag{4}$$

The local stress at the crack-tip can then be computed using the near-field stress and the stress intensity factor as

$$K = K_I + iK_{II} \tag{5}$$

$$\sigma = \frac{Re\left(Ka^{i\epsilon}\right)}{\sqrt{2\pi l}}, \ \tau = \frac{Im\left(Ka^{i\epsilon}\right)}{\sqrt{2\pi l}} \tag{6}$$

where a is the crack length, $\epsilon = \frac{1}{2\pi}ln\left(\frac{1-\beta}{1+\beta}\right)$, $\beta = \frac{1}{2}\frac{\mu_1(1-2v_2)-\mu_2(1-2v_1)}{\mu_1(1-v_2)+\mu_2(1-v_1)}$ are the materials mismatch parameters [24,25], μ_i, v_i are the shear modulus and Poisson's ratio for Cu and Si, respectively, where $i = 1, 2$, 1 represents Cu, 2 represents Si. $l = 100$ nm is the length scale for the investigated problem. The stress intensity factors were obtained using the LEFM FEA analysis.

The mesh details for the LEFM finite element analysis are shown in Figure 7a, where the plain strain 4-node bilinear quadrilateral elements were used in the region away from the crack-tip. The size of FE meshes was chosen to be less than 1/3 of pre-notch length, which was set as 100 nanometers. The mesh configuration used in this mode provided four contour integral paths to calculate J-integrals. The singular elements were then used near the crack-tip with a square root singularity [26]. The normal, shear stress and strain contours of the analyzed micro-cantilever beams are shown in Figure 7b. The analysis was then conducted for the six types of specimens with given tested failure loads (*P*) and geometrical characteristics. The phase angle was defined in terms of stress [25] as $\psi = \arctan\left(\frac{Im\left(Ka^{i\epsilon}\right)}{Re\left(Ka^{i\epsilon}\right)}\right)$ and plotted against the thickness-over-length ratio for the cantilever beams. The results (Table 2) showed that the variation in the beam height-over-length ratio provides a phase angle ranging from 16.7° to 83.7°, covering almost the half range of the mode-mix, ranging from 0 to +90 degrees. The normal and shear stress (σ_0, τ_0) obtained using Equation (6) at the failure load are then the mixed-mode fracture strength corresponding with the associated phase angle. The vectorial fracture strength can also be obtained by T = $\sqrt{\sigma_0^2 + \tau_0^2}$. The fracture toughness was also calculated using the critical stress intensity factors calculated following the equation [24] below,

$$\Gamma = \frac{(1 - \beta^2)}{E_*}\left(K_{Ic}^2 + K_{IIc}^2\right) \tag{7}$$

It should be noted here that these crack-tip stresses are essentially the stresses at *l* away from the crack-tip. The effect of the plastic zone was omitted, since the calculated stresses at these distances were much smaller than the yield strength of the Cu (216 MPa). However, the NLEFM analysis was nevertheless conducted to justify the negligence of the plastic effect.

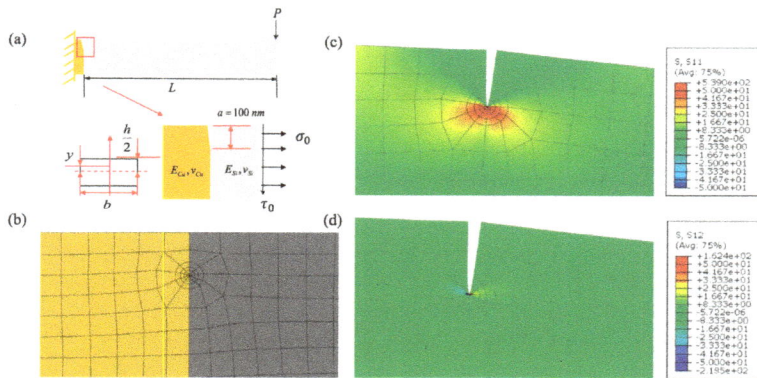

Figure 7. (**a**) Illustration of micro-cantilever beam with pre-crack, (**b**) FEA mesh details (yellow indicates Cu, grey indicates Si), (**c**) normal stress at the crack tip, (**d**) shear stress at the crack tip.

Table 2. Mixed-mode fracture analysis results.

L (μm)	ψ (Degree)	σ_0 (MPa)	τ_0 (MPa)	*T* (MPa)	Γ (J/m²)
1	83.7	10.5	25.6	27.7	5.7
4	66.0	20.7	18.0	27.4	5.6
8	48.4	21.8	16.0	27.0	5.4
12	36.9	25.0	10.0	26.9	5.4
18	26.6	26.3	5.0	26.8	5.3
30	16.7	26.7	0.0	26.7	5.3

(2) NLEFM Analysis

The non-linearity of the interfacial mixed-mode fracture typically comes from two perspectives: The cohesive behavior at the interface and the material's non-linearity. Based on our previous work, we concluded that the cohesive zone for the investigated Cu/Si interface was smaller than 100 nm.

Therefore, the cohesive zone analysis was not considered, since the cohesive zone length was much smaller than the characteristic length of the micro-cantilever beam. However, the material nonlinearity, in this case the Cu plasticity, had to be considered in the modeling to ensure the results obtained using LEFM were valid. In the NLEFM analysis, same geometrical characteristics and mesh configuration were used as in the LEFM. The only modification was the replacement of the elastic behavior of Cu with the measured elastic–plastic behavior from the micro-pillar experiment. All six types of specimens were modeled by applying the measured failure loads. The typical equivalent plastic strain contours are presented in Figure 8. The region where material has entered the plastic regime is labeled by the red dashed circles. The radius of these circles ranged from 10 to 15 nm, which were smaller than the 100 nm characteristic length scale used in the LEFM analysis, which validated the obtained mixed-mode fracture results.

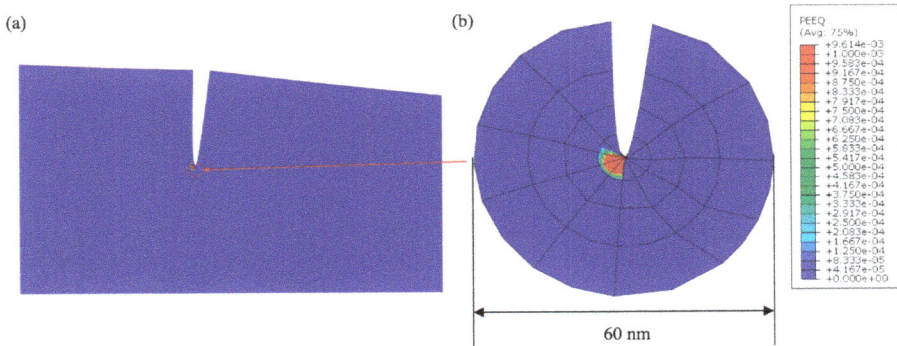

Figure 8. Non-linear fracture mechanical mode (NLEFM) analysis results of micro-cantilever beam for $L = 4$ μm: (**a**) Far-field view, (**b**) localized view near crack-tip showing equivalent plastic strain contour.

5. Results and Discussions

5.1. Strain Hardening of Cu Via

The yield strength measured from the micro-pillar experiment was close to those measured at the bulk scale. However, the Ramberg–Osgood parameter ($n = 3$) measured at micro-scale was much less than those typically measured at bulk scale ($n = 5$), which indicated a possible size effect caused by the reduced relative grain size. The average grain sizes measured for the TSV used in this study was about 500 nm [8], which was slightly smaller than the typical grain size observed at the bulk scale. The smaller grain size increased the total grain boundary area that contributed to the strain hardening mechanism, as illustrated by Taylor's theory [27–31]. This increased strain hardening behavior of Cu can effectively "lock" the plastic strain development within a small region, as observed in shear fracture of our previous work as well as in the micro-cantilever beam experiment. Therefore, it is worth noting here that the Cu plasticity had limited effects on the interfacial fracture of Cu/Si interface. The NLEFM results also confirmed that the crack-tip induced stress singularity caused a limited plastic effect. This however, was constrained within an area smaller than the characteristic length of the investigate interface. This constrain was also related to the limitation on the mode-mix, induced by varying the length-over-height ratio of the micro-cantilever beam. The pure mode-I and mode-II cases were not fully achieved, though closely approximated, avoided the growth of the plastic zone in the Cu via.

5.2. Mixed-Mode Cu/Si Interfacial Behavior

The phase angle versus the beam length (L) is plotted in Figure 9a. A decreasing trend was observed as the beam length increased. The range of the phase angle was from 16.7 to 83.7

degrees, covering most parts of the positive mode-mix (0–90°), which indicated completeness of the experimental data set in determining the fracture criterion at the Cu/Si interface.

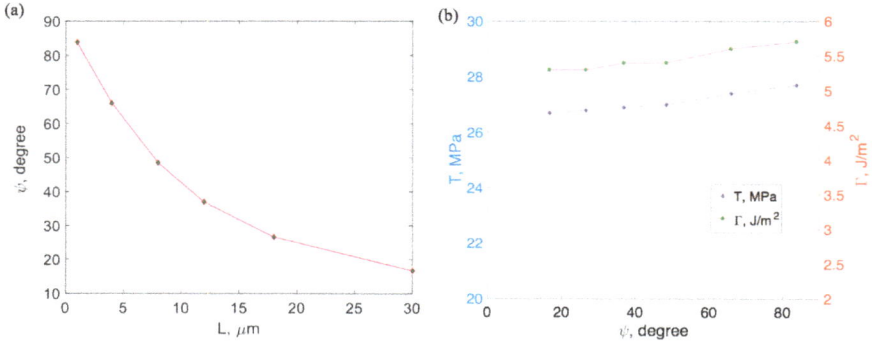

Figure 9. (**a**) Phase angle versus beam length, (**b**) vectorial fracture strength and toughness versus phase angle.

The mode-mix (in terms of phase angle) versus the vectorial interfacial strength (T) and the interfacial toughness (Γ) is shown in Figure 9b. A slight increase was observed for both values as the phase angle increased (i.e., more shear contribution is present). However, both the strength and toughness are relative, independent of the mode-mix. The average vectorial mixed-mode strength was found to be less dependent on the mode-mix. The average failure strength ($|T|$) was about 27 MPa, which was much lower than the yield strength of Cu (216 MPa) and the fracture strength of Si. Therefore, we suspected the liner materials at the Cu/Si interface contributed to this low interfacial strength.

Given these results, a fracture criterion was then proposed for the tested Cu/Si interface. Following the power-law failure criterion proposed by Carlsson et al. [15], the failure strength of the Cu/Si interface can be described as the following equation.

$$\left(\frac{\sigma_0}{\sigma_c}\right)^\lambda + \left(\frac{\tau_0}{\tau_c}\right)^\lambda = 1 \tag{8}$$

where σ_c, τ_c are the fracture strengths for pure mode-I (normal) and mode-II (shear) and λ is a fitting parameter, which was set at 1.8. The measured experimental data are then fitted with the proposed failure criterion, as shown in Figure 10. The dashed blue line shows the fitting of experimental data with $\sigma_c = \tau_c = T_{avg}$. The red solid line shows the fitting of $\sigma_c = 28$ MPa, $\tau_c = 26.5$ MPa. The better fitting of the experimental was observed when setting different fracture strengths for pure mode-I and mode-II. These results indicated that the mode-mix still had a moderate effect on the fracture strength, although the vectorial value stayed almost constant. When the combined normal and shear stress in the plane satisfy the condition of $\left(\frac{\sigma}{\sigma_c}\right)^{1.8} + \left(\frac{\tau}{\tau_c}\right)^{1.8} < 1$, the fracture initiation was not likely to occur.

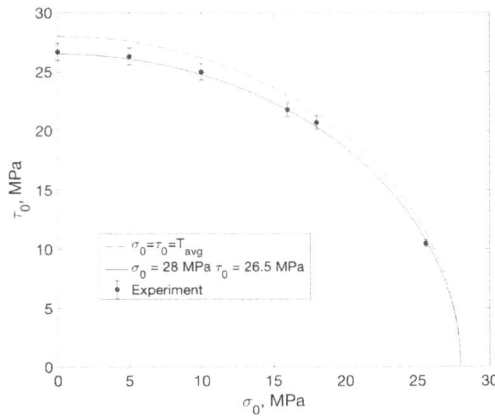

Figure 10. Failure criterion and experimental data.

6. Conclusions

This work combines the micro-pillar compression and micro-cantilever experiments to extract the mixed-mode fracture strength of the Cu/Si interface at the small scale (<100 nm). A series of micro-cantilever beam specimens with various beam lengths were fabricated and tested covering almost the full range of the positive mode-mix (0–90°). The following conclusions were drawn from the experiment and analysis results:

(1) The mixed-mode fracture at the Cu/Si interface is brittle in nature. No significant cohesive zone was observed, nor does the plastic hinge that is typically found for the micro-cantilever beam consist of metal materials.

(2) The effect of the plastic behavior of the Cu was negligible. The NLEFM analysis results showed that this is mostly due to the constrained plastic region, which was much smaller than the characteristic length (100 nm).

(3) The vectorial fracture strength and toughness were obtained from the analysis, which indicated that the effect of the mode-mix was limited, which was suspected to be related to the constrained plastic zone.

(4) A power-law fracture strength criterion could be used to fit the experimental data with close agreement. The criterion could be used for the design and engineering of TSV structures in combined loadings.

Future studies will focus on the evaluation of the effects of residual stress, interfacial defects, and impurities of Cu on the mixed-mode interfacial strengths. In particular, the residual stress built-up at the Cu/Si interface, due to elastic and coefficient of thermal expansion (CTE) mismatch, will cause deviations on the extracted mixed-mode fracture strengths measured. Under extreme circumstances, plastic zones can be formed at the interface, as well providing the possible reduction in the mixed-mode fracture strengths. The residual stress generated during the fabrication process will be estimated and taken into consideration in future studies.

Author Contributions: C.W. (Chenglin Wu) and Y.L. conceived and designed the experiments; C.W. (Chenglin Wu) performed the experiments; C.W. (Chenglin Wu) and C.W. (Congjie Wei) analyzed the data; C.W. (Chenglin Wu) contributed reagents/materials/analysis tools; C.W. (Chenglin Wu) and C.W. (Congjie Wei) wrote the paper.

Funding: The author would like to thank the seed funding from Material Research Center for this project.

Conflicts of Interest: The authors declare no conflict of interest.

References

1. Garrou, P.; Bower, C.; Ramm, P. *Handbook of 3D Integration, Volume 1: Technology and Applications of 3D Integrated Circuits*; John Wiley & Son: Hoboken, NJ, USA, 2011.
2. Jiang, T.; Im, J.; Huang, R.; Ho, P.S. Through-silicon via stress characteristics and reliability impact on 3D integrated circuits. *MRS Bull.* **2015**, *40*, 248–256.
3. De Wolf, I.; Croes, K.; Pedriera, O.V.; Labie, R.; Redolfi, A.; Van De Peer, M.; Vanstreels, K.; Oroko, C.; Vandevelde, B.; Beyne, E. Cu pumping in TSVs: Effect of pre-CMP thermal budget. *Microelectron. Reliab.* **2011**, *51*, 1856–1859. [CrossRef]
4. Heryanto, A.; Punta, W.N.; Trigg, A.; Gao, S.; Kwon, W.S.; Che, F.X.; Ang, F.X.; Wei, J.; Made, R.I.; Gan, C.L.; et al. Effect of copper TSV annealing on via protrusion for TSV wafer fabrication. *J. Electron. Mater.* **2012**, *41*, 2533–2542. [CrossRef]
5. De Messemaeker, J.; Pedreira, O.V.; Vandevelde, B.; Philipsen, H.; De Wolf, I.; Beyne, E.; Croes, K. Impact of post-plating anneal and through-silicon via dimensions on Cu pumping. In Proceedings of the 2013 IEEE 63rd Electronic Components and Technology Conference, Las Vegas, NV, USA, 28–31 May 2013.
6. Jiang, T.; Wu, C.; Im, J.; Huang, R.; Ho, P.S. Impact of grain structure and material properties on via extrusion in 3D interconnects. *J. Microelectron. Electron Pack.* **2015**, *12*, 118–122. [CrossRef]
7. Jiang, T.; Wu, C.; Spinella, L.; Im, J.; Tamura, N.; Kunz, M.; Son, H.-Y.; Kim, B.G.; Huang, R.; Ho, P.S. Plasticity mechanism for copper extrusion in through-silicon vias for three-dimensional interconnects. *Appl. Phys. Lett.* **2013**, *103*, 211906. [CrossRef]
8. Wu, C.; Huang, R.; Leitchi, K.M. Characterizing Interfacial Sliding of Through-Silicon-Via by Nano-Indentation. *IEEE Trans. Device Mater. Relaib.* **2017**, *17*, 355–363. [CrossRef]
9. Wu, C.; Jiang, T.; Im, J.; Leichti, K.M.; Huang, R.; Ho, P.S. Material characterization and failure analysis of through-silicon vias. In Proceedings of the 21th International Symposium on the Physical and Failure Analysis of Integrated Circuits (IPFA), Marina Bay Sands, Singapore, 30 June–4 July 2014.
10. Wu, C.; Taghvaee, T.; Wei, C.; Ghasemi, A.; Chen, G.; Leventis, N.; Gao, W. Multi-scale progressive failure mechanism and mechanical properties of nanofibrous polyurea aerogels. *Soft Matter.* **2018**, *14*, 7801–7808. [CrossRef]
11. Li, Y.; Liao, Y.W.; Taghvaee, T.; Wu, C.; Ma, H.; Leventis, N. Bioinspired Strong Nanocellular Composite Prepared with Magnesium Phosphate Cement and Polyurea Aerogel. *Mater. Lett.* **2019**, *237*, 274–277. [CrossRef]
12. Beegan, D.; Chowdhury, S.; Laugier, M.T. A nanoindentation study of copper films on oxidised silicon substrates. *Surf. Coat. Technol.* **2003**, *176*, 124–130. [CrossRef]
13. Beegan, D.; Chowdhury, S.; Laugier, M.T. Comparison between nanoindentation and scratch test hardness (scratch hardness) values of copper thin films on oxidised silicon substrates. *Surf. Coat. Technol.* **2007**, *201*, 5804–5808. [CrossRef]
14. Fang, T.-H.; Chang, W.-J. Nanomechanical properties of copper thin films on different substrates using the nanoindentation technique. *Microelectron. Eng.* **2003**, *65*, 231–238. [CrossRef]
15. Carlsson, L.; Gillespie, J., Jr.; Pipes, R. On the analysis and design of the end notched flexure (ENF) specimen for mode II testing. *J. Compos. Mter.* **1986**, *20*, 594–604. [CrossRef]
16. Chai, H. Experimental evaluation of mixed-mode fracture in adhesive bonds. *Exp. Mech.* **1992**, *32*, 296–303. [CrossRef]
17. Dollhofer, J.; Beckert, W.; Lauke, B.; Schneider, K. Fracture mechanics characterization of mixed-mode toughness of thermoplast/glass interfaces (brittle/ductile interfacial mixed-mode fracture). *J. Adhesion Sci. Technol.* **2001**, *15*, 1559–1587. [CrossRef]
18. Kfouri, A.; Brown, M. A fracture criterion for cracks under mixed-mode loading. *Fatigue Fract. Eng. Mater. Struct.* **1995**, *18*, 959–969. [CrossRef]
19. Wu, C. Using Far-Field Measurements for Determining Mixed-Mode Interactions at Interfaces. Ph.D. Thesis, The University of Texas at Austin, Austin, TX, USA, 2017.
20. Ramberg, W.; Osgood, W.R. *Description of Stress-Strain Curves by Three Parameters*; Technical note; National Advisory Committee for Aeronautics: Washington, DC, USA, 1943.
21. Zhang, C.; Leng, Y.; Chen, J. Elastic and plastic behavior of plasma-sprayed hydroxyapatite coatings on a Ti–6Al–4 V substrate. *Biomaterials* **2001**, *22*, 1357–1363. [CrossRef]

22. Lin, D.; Dimitriadis, E.; Horkay, F. Elasticity of rubber-like materials measured by AFM nanoindentation. *Express Polym. Lett.* **2007**, *1*, 576–584. [CrossRef]

23. Albrecht, J.; Weissbach, M.; Auersperg, J.; Rzepka, S. Method for assessing the delamination risk in BEoL stacks around copper TSV applying nanoindentation and finite element simulation. In Proceedings of the 2017 IEEE 19th Electronics Packaging Technology Conference (EPTC), Singapore, 6–9 December 2017.

24. Dundurs, J. Discussion of edge-bonded dissimilar orthogonal elastic wedges under normal and shear loading. *J. Appl. Mech.* **1969**, *36*, 650–652. [CrossRef]

25. Hutchinson, J.W.; Suo, Z. Mixed mode cracking in layered materials. In *Advances in Applied Mechanic*; Elsevier: Amsterdam, The Netherlands, 1991; pp. 63–191.

26. Aoki, S.; Kishimoto, K.; Sakata, M. Crack-tip stress and strain singularity in thermally loaded elastic-plastic material. *J. Appl. Mech.* **1981**, *48*, 428–429. [CrossRef]

27. Kocks, U.; Mecking, H. Physics and phenomenology of strain hardening: The FCC case. *Mater. Sci.* **2003**, *48*, 171–273. [CrossRef]

28. International Series on the Strength and Fracture of Materials and Structures. In *International Series on the Strength and Fracture of Materials and Structures, Proceedings of the 8th International Conference on the Strength of Metals and Alloys, Tampere, Finland, 22–26 August 1988*; Haasen, P., Gerold, V., Kostorz, G., Eds.; Pergamon: Oxford, UK, 1988; p. ii.

29. Mecking, H.; Kocks, U.; Hartig, C. Taylor factors in materials with many deformation modes. *Scripta Mater.* **1996**, *35*. [CrossRef]

30. Kocks, U. The relation between polycrystal deformation and single-crystal deformation. *Metall. Mater Trans. B* **1970**, *1*, 1121–1143. [CrossRef]

31. Tomé, C.; Kocks, U. The yield surface of hcp crystals. *Acta Metall.* **1985**, *33*, 603–621. [CrossRef]

micromachines

MDPI

Article

Testing Effects on Shear Transformation Zone Size of Metallic Glassy Films Under Nanoindentation

Yi Ma [1],*, Yuxuan Song [1], Xianwei Huang [1], Zhongli Chen [2] and Taihua Zhang [1,3],*

[1] College of Mechanical Engineering, Zhejiang University of Technology, Hangzhou 310014, China;
    1111702002@zjut.edu.cn (Y.S.); huangxw@zjut.edu.cn (X.H.)
[2] China Jiliang University, Hangzhou 310018, China; chenzhongli@cjlu.edu.cn
[3] Institute of Solid Mechanics, Beihang University, Beijing 100191, China
* Correspondence: may@zjut.edu.cn (Y.M.); zhangth66@buaa.edu.cn (T.Z.);
    Tel.: +86-571-8832-0132 (Y.M. & T.Z.)

Received: 22 October 2018; Accepted: 28 November 2018; Published: 30 November 2018

check for updates

Abstract: Room-temperature creep tests are performed at the plastic regions of two different metallic glassy films under Berkovich nanoindetation. Relying on the strain rate sensitivity of the steady-state creep curve, shear transformation zone (STZ) size is estimated based on the cooperative shear model (CSM). By applying various indentation depths, loading rates, and holding times, the testing effects on the STZ size of metallic glasses are systematically studied. Experimental results indicate that STZ size is greatly increased with increased indentation depth and shortened holding time. Meanwhile, STZ size is weakly dependent on the loading history. Both the intrinsic and extrinsic reasons are discussed, to reveal the testing effects on the nanoindentation creep flow and STZ size.

Keywords: metallic glass; nanoindentation; creep; strain rate sensitivity; shear transformation zone

1. Introduction

Metallic glasses are at the cutting edge of new-structure material research and have great potential to be utilized as engineering materials for their excellent mechanical properties [1–4]. Due to its unique atomic configuration, i.e., non-crystalline but with a short-range order structure, metallic glass is also an important part of condensed matter physics. However, localized shear banding dominates in plastic deformation and catastrophic failure always occurs under tension [5]. A great deal of research efforts have focused on revealing the mechanism of plastic deformation and improving workability in metallic glasses. Owing to the original work of Argon [6], the deformation unit with a local rearrangement of atoms, also referred to as the shear transformation zone (STZ), has been widely applied to analyze the low-temperature deformation of metallic glasses. Being different from a structural defect, STZ is defined by its transience, i.e., it can only be identified from the atomic structures before and after deformation. The details of STZ evolution are mostly studied using computer simulations, in relation to its shape, configuration, and activation mechanism [7–10].

In recent years, following the cooperative shear model (CSM) by Johnson and Samwer [11], Pan et al. developed an experimental method to estimate STZ volume relying on strain rate sensitivity (SRS) through rate-jump nanoindentation [12–14]. In Pan's work, STZ size displayed a strong correlation with Poisson's ratio and structure state [12,15], and could be closely tied to the ductility of bulk metallic glasses. However, the experimental results determined using the rate-jump method are in doubt. Bhattacharyya et al. revealed that pile-up in nanoindentation would significantly affect the SRS values using the rate-jump method [16], hence the questionable STZ volume. Moreover, strain rate sensitivities in some kinds of metallic glasses are extremely low, and even negative. Thus, the measuring error would be significant for calculating SRS values and the corresponding result concerning STZ size might be unreliable.

Indentation creep has been the most extended method for studying strain rate sensitivity [17]. The holding stage during nanoindentation could be much more time-saving in comparison to the traditional creep test, due to its high accuracy for recording creep displacement [18,19]. Nanoindentation creep investigations have been widely performed in small-scale nanocrystalline materials [20–23]. In the authors' previous work, spherical nanoindentation creep behaviors in metallic glassy films with various compositions were carefully studied [24]. Furthermore, STZ sizes and their correlations with intrinsic parameters of metallic glasses such as sample dimension [25,26], structure state [27,28], residual stress [29], and glass transition temperature (T_g) [24], were obtained, relying on the SRS of creep curves. It has been recognized that nanoindentation creep behaviors are dependent on testing conditions in both crystalline and amorphous alloys [30–36]. Now, one question naturally arises: Does the STZ size of metallic glass also depend on extrinsic testing effects? With this in mind, two distinct metallic glassy films are prepared, which could provide a clean and smooth surface for nanoindentation. In the present study, the testing effects of indentation depth, loading rate, and holding time on STZ size are studied using the creep method.

2. Materials and Methods

Zr-Cu-Ni-Al and La-Co-Al films were deposited on a silicon wafer in a DC magnetron sputtering system at room temperature in pure argon gas. The 2-inch target alloys adopted in the chamber were $Zr_{64}Cu_{16}Ni_{10}Al_{10}$ and $La_{60}Co_{20}Al_{20}$, at.%, which was prepared from high-purity (99.99 wt.%) elements using vacuum casting. The target was installed at the bottom while the silicon wafer was stuck onto the sample platform, which was right above the target. The target-to-substrate distance was kept constant, equal to 100 mm. The base pressure of the chamber was kept at about 5×10^{-7} Torr before deposition and the working argon pressure was set at about 1 mTorr. The power on the target was fixed at 120 W during the deposition and the sputtering time was two hours. The film thickness could be directly measured from the cross-section using a scanning electron micrograph (SEM). By means of an X-ray energy dispersive spectrometer (EDS) attached to the SEM, the chemical compositions of as-prepared films were detected as $Zr_{55}Cu_{15}Ni_{13}Al_{17}$ and $La_{55}Co_{20}Al_{25}$, respectively. The amorphous nature was confirmed using X-ray diffraction (XRD) with Cu K_α radiation [24].

Nanoindentation experiments were conducted at a constant temperature of 20 °C on an Agilent Nano Indenter G200 with the dynamic contact module (DCM), by which higher resolution in both force and displacement and less sensitivity to the environment could be attained. The constant temperature was controlled by the air condition. A standard Berkovich indenter was applied, the tip of which was detected to be perfect using transmission electron microscopy (TEM), to avoid the tip bluntness effect on the mechanical response [37,38]. The constant load-holding method was used in this study to explore the creep flows and strain rate sensitivities under various testing conditions. The indenter was held for 500 s at various depths of 50 nm, 100 nm, 200 nm, and 350 nm (only for La-Co-Al film), with a constant loading rate of 0.2 mN/s. At a constant holding depth of 200 nm, the loading rate effect on the creep flow was studied; during these tests, four different loading rates of 0.035 mN/s, 0.075 mN/s, 0.2 mN/s, and 0.75 mN/s were employed. Furthermore, five different holding durations of 15 s, 50 s, 100 s, 500 s, and 1000 s were adopted to study the holding time effect on the value of strain rate sensitivity and STZ size. All the above-mentioned testing parameters were carefully chosen, in view of the potential disturbances of the substrate effect, tip bluntness effect, and the influences of thermal drift and instrument error on experimental results. The creep tests were launched until thermal drift was reduced to below 0.02 nm/s. Meanwhile, drift correction, which was calibrated at 10% of the maximum load during the unloading process, was strictly performed. To ensure the reliability of the creep results, twenty nanoindentation measurements were conducted for each test.

3. Results and Discussion

Figure 1 shows the film cross-sections of both samples; the thicknesses of Cu-Zr-Ni-Al and La-Co-Al films could be directly measured as 1.6 µm and 2.7 µm, respectively. Figure 2 shows the typical load-displacement (*P-h*) curves of 500-s creep tests with a holding depth of 200 nm and a loading rate of 0.2 mN/s. The room-temperature creep deformation could be clearly observed at the holding stage. At the onset of the holding stage, severe plastic deformation beneath the Berkovich indenter had already occurred. This was, intrinsically, why creep deformation could easily occur even in many high-melting point materials under nanoindentation at room temperature. The creep flows of metallic glassy films could be more intuitively recognized from the relationship between the creep displacements and holding time, as plotted in Figure 3a. The creep curves could be divided into two distinct stages, as transient creep and steady-state creep. The approximate critical point for the creep transition is marked with an arrow in Figure 3a for both samples. In comparison, the transient creep stage was much shorter in La-Co-Al than in Zr-Cu-Ni-Al. At the transient stage, the creep displacement increased relatively fast, but the creep rate dropped rapidly. Then, the creep displacement turned out to be slow and increased almost linearly with time at the steady-state stage. Clearly, creep deformation is more pronounced in La-Co-Al than in Zr-Cu-Ni-Al, which could be expected due to the lower glass transition temperature (T_g) and hardness. The creep curves for both samples can be perfectly fitted ($R^2 > 0.99$) using an empirical law [39]:

$$h(t) = h_0 + a(t - t_0)^b + kt \tag{1}$$

where h_0 and t_0 are the displacement and time at the beginning of the holding stage, and a, b, k are the fitting constants.

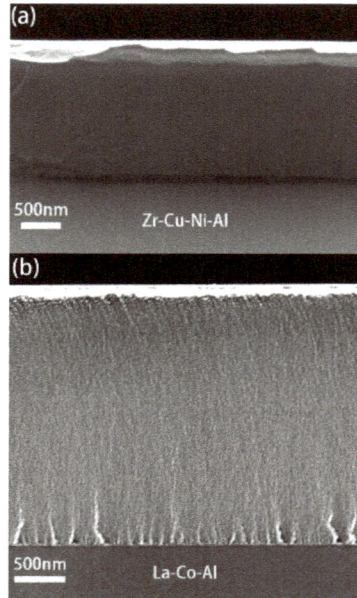

Figure 1. Typical cross-sections of: (**a**) Zr-Cu-Ni-Al; and (**b**) La-Co-Al films using scanning electron micrograph (SEM), film thickness could be measured.

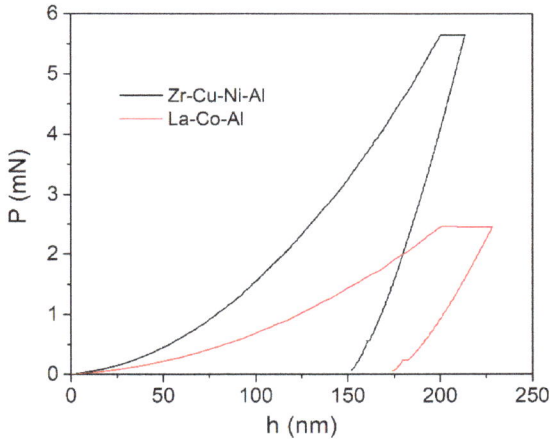

Figure 2. Representative load-displacement curves of the 500-s holding creep tests with a maximum displacement of 200 nm for Zr-Cu-Ni-Al and La-Co-Al films.

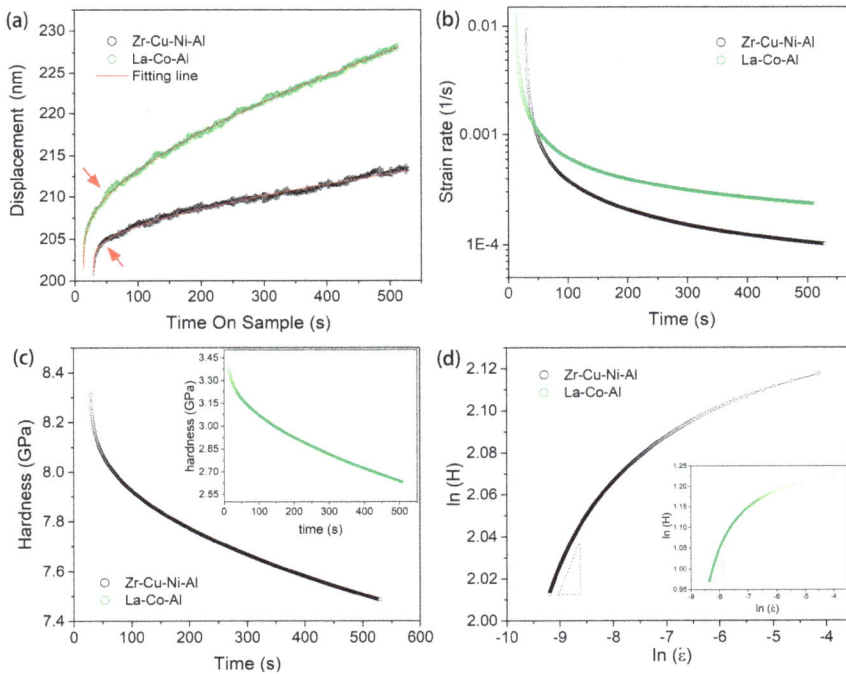

Figure 3. (a) the creep displacements during the holding stage versus holding time, which can be perfectly fitted using an empirical law, and the critical point of creep transition from transient stage to steady-state stage is marked with an arrow; (b) the creep strain rate versus holding time; (c) the hardness versus holding time; (d) the log-log correlation between the hardness and strain rate obtained from the creep, strain rate sensitivity can be thus estimated by linear fitting of the steady-state part.

The value of the SRS exponent m can be evaluated via [40]:

$$m = \frac{\partial \ln H}{\partial \ln \dot{\varepsilon}} \tag{2}$$

In the present Berkovich nanoindentation process, the strain rate during the holding stage can be calculated as $\dot{\varepsilon} = \frac{1}{h_p}\frac{dh_p}{dt}$ and the hardness is defined as $H = \frac{P}{Ch_p^2}$. C is the tip area coefficient and is rectified upon testing on the standard fused silica, equal to 23.6 here. h_p is the contact depth and could be estimated as $h_p = h_i - \varepsilon P/S$, where h_i is the total nanoindentation displacement, $\varepsilon = 0.72$ for a Berkovich tip, and S is the stiffness deduced from the unloading curve. Based on the fitting line of the creep curve in Figure 3a, the variations in the strain rate and hardness as a function of the holding time are obtained, as shown in Figure 3b,c, respectively. Figure 3d shows the log-log correlation between the hardness and strain rate during the holding stage. Accordingly, m can be obtained by linearly fitting the part of the steady-state creep (here, we adopted the last 200-s holding part). For reliability, nine effective creep curves were employed to reach an average value of SRS for each sample.

The STZ volume can then be estimated accordingly using the cooperative shear model (CSM) of Johnson and Samwer [11], based on the SRS obtained from the creep. In the CSM model, the activation energy of the STZ is defined as:

$$W_{STZ} = 4R_0 G_0 \gamma_C^2 (1 - \tau/\tau_0)^{\frac{3}{2}} \xi \Omega \tag{3}$$

Thus, the correlation between the STZ volume Ω and the activation volume V^* can be obtained through directly differentiation of the activation energy W_{STZ}, given by:

$$\Omega = \frac{\tau_0}{6R_0 G_0 \gamma_C^2 (1 - \tau/\tau_0)^{\frac{1}{2}} \xi} V^* \tag{4}$$

where $R_0 \approx 1/4$ and $\xi \approx 3$ are constants, τ and τ_0 are threshold shear resistances at temperature T and 0 K, G_0 is the shear modulus at 0 K, the average elastic limit $\gamma_C \approx 0.027$, $\tau_0/G = 0.036$, and the value of τ/τ_0 can be estimated using the constitutive equation:

$$\tau/G = \gamma_{C0} - \gamma_{C1}(T/T_g)^{2/3} \tag{5}$$

where $\gamma_{C0} = 0.036 \pm 0.002$, $\gamma_{C1} = 0.016 \pm 0.002$, the shear modulus $G \approx \frac{E}{2(1+v)}$ has a weak temperature dependency for a metallic glass [11]. The STZ activation volume V^* can be expressed as [40]:

$$V^* = \frac{kT}{m\tau_y} \tag{6}$$

Here, τ_y is the critical shear stress upon the traditional tensile or compressive tests, with an empirical correlation of $\tau_y \approx H/3\sqrt{3}$. The hardness value at the initial holding stage was adopted to estimate the flow stress. Once the STZ activation volume V^* had been determined, STZ activation energy and volume could be calculated using Formulas (3) and (4). According to the dense-packing hard-sphere model of metallic glass [41], wherein the average atomic radius $r \approx (\sum_i^n A_i r_i^3)^{1/3}$, A_i and r_i are the atomic fraction and atomic radius of each element, respectively, the atoms contained in an STZ could be estimated.

3.1. Indentation Size Effect

Figure 4a,b shows the typical creep flows at various holding depths for Zr-Cu-Ni-Al and La-Co-Al, respectively. It should be noted that a load-holding test was not conducted at 350 nm for the Zr-Cu-Ni-Al film due to the potential substrate effect [42–44]. In order to recognize the indentation size effect on creep behavior more directly, the starting points (including both the holding time and creep displacement) for all the creep curves were set to be zero. Clearly, the creep flow was enhanced by increasing the initial holding depth. Furthermore, all the creep curves could be perfectly fitted using Equation (1). The total creep displacement at each holding depth was recorded for both samples, as shown in Figure 5a. The enhancement of the creep displacement by increasing the holding depth was more evident in La-Co-Al than in Zr-Cu-Ni-Al. From the perspective of structure agitation under nanoindentation, it is qualitatively claimed that the free volume content of metallic glass would be increased with pressed depth, hence promoting creep flow at a larger holding depth. In addition, the shear band density and excess free volume generated during nanoindentation would be composition-dependent in metallic glasses [45], which could be the reason for the different performances in relation to the indentation size effect in Zr-Cu-Ni-Al and La-Co-Al. Such indentation size effect on creep deformation is commonly observed in nanoindentation creep, though its mechanism is complicated. For a spherical indenter, the plastic deformation beneath the indenter would be more severe as the pressed depth increases. More excess free volume would be generated in the plastic zone, causing improved atomic mobility. It could be conceivable that a larger holding depth would facilitate creep flow (creep strain) under spherical nanoindentation, while for a standard Berkovich indenter, the imposed plastic strain and stress distribution during nanoindentation are self-similar at various pressed depths. Theoretically, creep displacement would be proportional to the initial holding depth and creep strain would be invariable under Berkovich indentation. Practically, the stress distribution beneath the indenter is much more complicated than that in uniaxial testing and creep deformation could not occur uniformly around the plastic region. Here, we define creep strain as $\Delta h/h_c$, where Δh is the total creep displacement and h_c is the contact displacement at the beginning of the holding stage. The correlations between creep strain and initial holding depth are shown in Figure 5b for both samples. The creep strain was gradually decreased with the increased initial holding depth. This phenomenon has been revealed in crystalline/amorphous nanolaminates, where the apparatus error of "indenter overshoot" plays an important role at the very beginning of the holding stage [46]. Figure 5c,d exhibits the hardness values at the beginning of the holding stage, which apparently decrease with increasing pressed depth.

The SRS at various holding depths are computed using steady-state creep curves for both samples. The obtained values are in the order of 10^{-1}, which is remarkably higher than those reported using the rate-jump method. The rate-jump method is conducted under quasi-static loading and suffered instantaneous plastic deformation, the results of which are always a magnitude less than those using creep. Figure 6a,b clearly shows that SRS decreased with the increasing holding depth. Based on the above values of SRS and hardness, the STZ volume and size of both samples could be determined. As depicted in Figure 6c,d, we can conclude that STZ size increases with increasing holding depth. The calculated values of STZ size increased from 10 atoms to 53 atoms for Zr-Cu-Ni-Al, and from 9 atoms to 46 atoms for La-Co-Al, as the holding depths increased from 50 nm to 200 nm and 50 nm to 350 nm, respectively. The STZ volume and size were, respectively, in the range of 0.1~1 nm^3 and 10~60 atoms, which corresponds well with previous studies on metallic glasses detected using experimental methods and molecular dynamic simulation [47–49]. On the other hand, the STZ size of La-Co-Al is smaller than Zr-Cu-Ni-Al at the same holding depth, which could be due to its lower T_g~500 K (730 K for Zr-Cu-Ni-Al) [50]. The glass transition temperature is a vital thermodynamic parameter of metallic glass, which is closely connected to the atomic structure and plastic deformation [51].

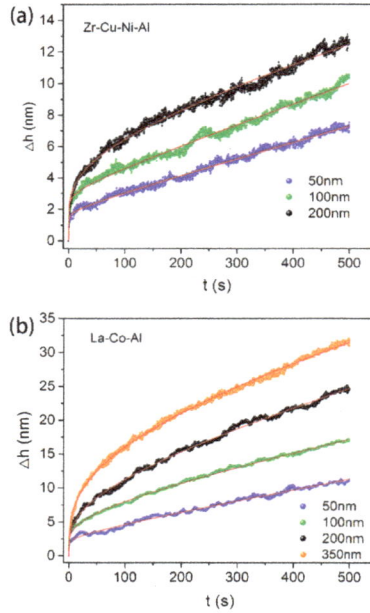

Figure 4. Typical creep displacements versus holding time at various initial holding depths for (**a**) Zr-Cu-Ni-Al and (**b**) La-Co-Al.

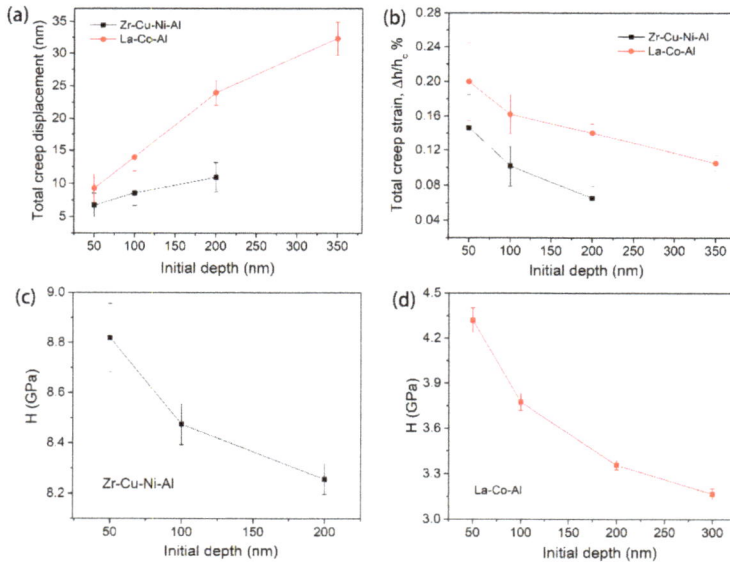

Figure 5. (**a**) The total creep displacement and (**b**) creep strain for at various initial holding depths for both samples; (**c**) hardness at the onset of the holding stage as a function of initial holding depth for (**c**) Zr-Cu-Ni-Al and (**d**) La-Co-Al.

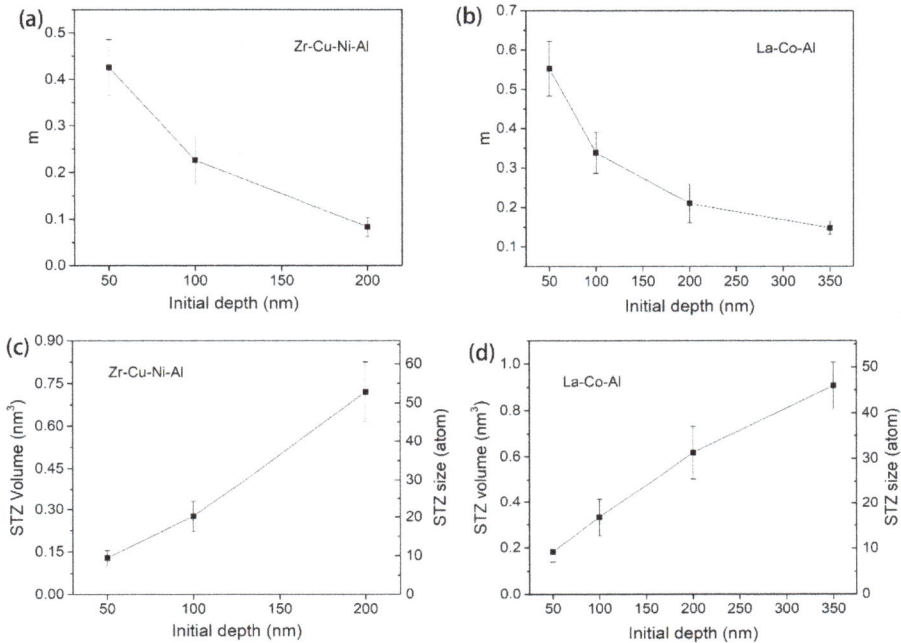

Figure 6. (**a**,**b**) Strain rate sensitivity, and (**c**,**d**) shear transformation zone (STZ) size as a function of initial holding depth for Zr-Cu-Ni-Al and La-Co-Al.

The indentation size effect on creep deformation and strain rate sensitivity (or stress exponent) has been widely reported in both crystalline and amorphous alloys [30–33]. However, the correlation between the activation volume of the plastic unit and pressed depth is rarely investigated. In Jang et al.'s work [49], it was reported that the STZ size of a Zr-based metallic glass was unchanged under spherical indenters with different radius using the statistical method. However, the initial conditions such as stress distribution and plastic deformation in their work were changed with different indenters, which significantly influenced the emergence of the first pop-in event. Strictly speaking, this was merely the indenter radius effect rather than the indentation size effect on STZ size. What is more, the indentation size effect on STZ size cannot be fully studied using the statistical method or rate-jump method in nanoindentation, due to their testing principles. To the authors' best knowledge, the results presented here are the first report revealing the indentation size-dependent STZ size in metallic glasses.

3.2. Loading Rate Effect

Figure 7 shows the typical creep curves during the holding stage under four different loading rates. The initial holding depth is constant as 200 nm for both samples. A fast loading sequence is thought to facilitate nanoindentation creep deformation during the holding stage [34–36]. However, the loading rate effect presented here on the creep flow of metallic glass is composition-dependent. For the Zr-Cu-Ni-Al film in Figure 7a, the creep deformation increases weakly when increasing the loading rate from 0.035 mN/s to 0.2 mN/s, and proves much more pronounced under 0.75 mN/s. Qualitatively, the stimulate effect on creep deformation by a high loading rate is more remarkable at the transient stage (0~50 s) than the steady-state stage for the Zr-Cu-Ni-Al film. For the La-Co-Al film in Figure 7b, on the other hand, the creep flows are observed to be history-independent. There is no obvious distinction among the creep curves with various loading rates. Figure 8a,b summarizes the correlations between nanoindentation hardness and loading rate for Zr-Cu-Ni-Al and La-Co-Al, respectively. The indentation hardness is increased with the increasing loading rate (strain rate) for both samples, which confirms the positive strain rate sensitivity. In comparison, the growing rate of hardness on the loading rate is much smaller in La-Co-Al than in Zr-Cu-Ni-Al, i.e., hardness increases from about 3.32 GPa to 3.37 GPa for La-Co-Al (growing rate ~1.5%), and from 7.9 GPa to 8.3 GPa for Zr-Cu-Ni-Al (growing rate ~5%), as the loading rate increases from 0.035 mN/s to 0.75 mN/s. It is worth noting that the loading rate effect on hardness is consistent with that on creep flow for both samples, i.e., positive for Zr-Cu-Ni-Al and insensitive for La-Co-Al. The promoting effects of the loading rate on hardness and creep flow in Zr-Cu-Ni-Al are commonly observed in metallic glasses, which could be explained from the perspective of structure agitation and the generation of excess free volume. For the La-Co-Al film, on the other hand, such a tiny variation of hardness illustrates that the process of plastic deformation and the structure state at the peak load do not change as the loading rate increases. As a consequence, the creep behaviors of La-Co-Al would be insensitive to the variation of loading sequences. The computed values of strain rate sensitivity from steady-state creep curves are summarized in Figure 8c,d for Zr-Cu-Ni-Al and La-Co-Al, respectively. As the loading rate increases from 0.035 mN/s to 0.75 mN/s, the average value of m generally increases from 0.07 to 0.13 for Zr-Cu-Ni-Al whilst it lies in the range of 0.22~0.18 for La-Co-Al.

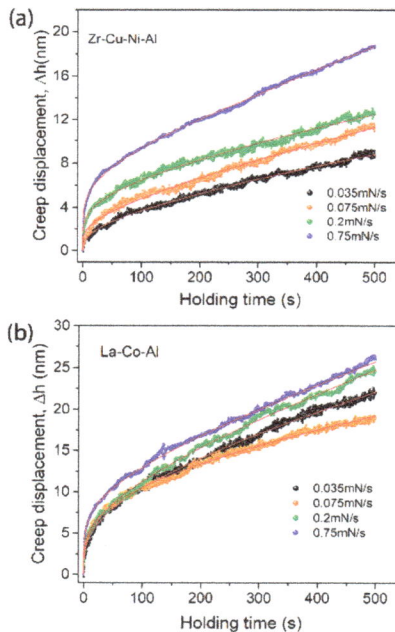

Figure 7. Typical creep displacements versus holding time with four different loading rates at the initial holding depth of 200 nm for (**a**) Zr-Cu-Ni-Al and (**b**) La-Co-Al; the creep curves could be perfectly fitted.

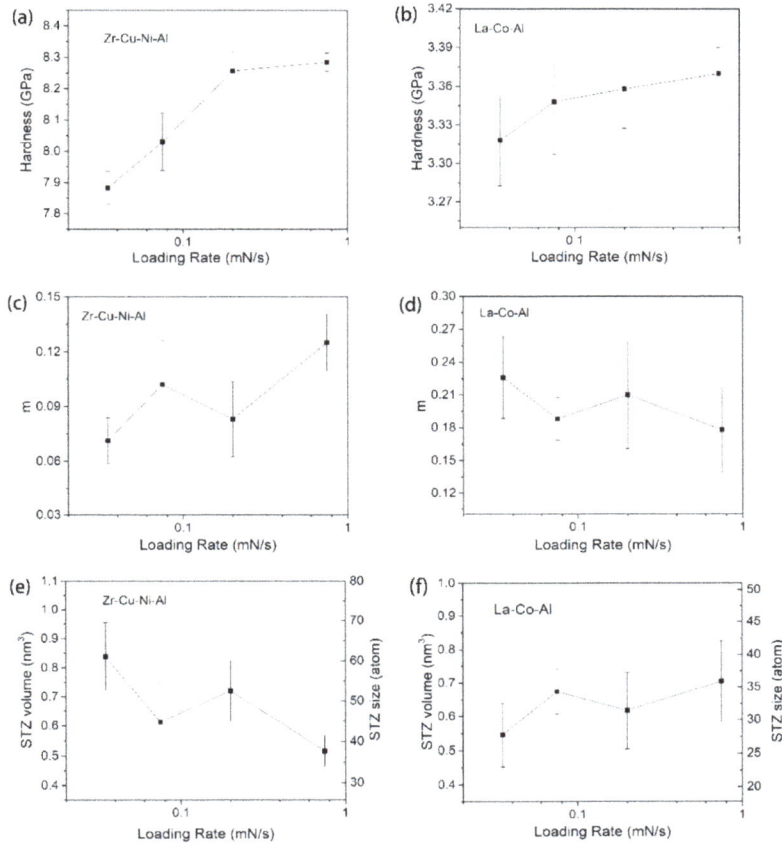

Figure 8. Hardness at the onset of the holding stage, strain rate sensitivity, and STZ size as a function of the loading rate for (**a,c,e**) Zr-Cu-Ni-Al and (**b,d,f**) La-Co-Al.

STZ volume and the atoms it contained were calculated for both samples, as listed in Figure 8e,f. For Zr-Cu-Ni-Al, STZ, the size is reduced to about 40 atoms under 0.75 mN/s and enlarged to 60 atoms under 0.035 mN/s. The results presented here could be connected to the loading rate-dependent creep behaviors. In Choi et al.'s work, it was also revealed that the STZ size of a Zr-Cu-Ni-Al-Ti metallic glass was decreased from 29 atoms to 17 atoms as the loading rate increased from 0.5 mN/s to 10 mN/s, using the nanoindentation statistical method which relies on the cumulative distribution of yield stress [52]. For La-Co-Al, on the other hand, the fluctuation of average STZ sizes with loading rate is quite small, at around 32 ± 2 atoms. Clearly, the loading rate effect on the STZ size of metallic glasses using the stress relaxation method is weak and composition-dependent. It was revealed that STZ size is closely connected to the free volume content [49], STZ size increases from 25 atoms to 33 atoms after sub-T_g annealing in a Zr-Cu-Ni-Al-Ti bulk metallic glass. As it is conceived that more excess free volume is generated beneath the indenter at a higher loading rate, a smaller STZ size can be expected at 0.75 mN/s for Zr-Cu-Ni-Al. For La-Co-Al, on the other hand, the free volume content might be insensitive to the loading rate, as mentioned above, hence the small variation in STZ size.

3.3. Holding Time Effect

As illustrated in Figure 3d, strain rate sensitivity is not a constant throughout the whole holding stage, even under fixed testing conditions. Clearly, the value of SRS m increases with the holding time. However, stiffness S is also changed with pressed depth, which is important for calculating the hardness and strain rate. Strictly speaking, we could not directly obtain the correlation between m and the holding time by differentiating the lnH-$ln\dot{\varepsilon}$ curve, in which S is fixed for simplicity. In order to ensure the accuracy of the holding time effect on STZ size, creep tests with various durations were performed. Figure 9 shows the hardness at the onset of the holding stage for both samples, as a function of the holding time. The decrease in hardness was due to the fact that S is increased with pressed depth (holding time). Figure 10a shows the correlation between strain rate sensitivity and holding time. In comparison to the result of the 500-s holding test, m is nearly doubled after 1000 s holding, i.e., 0.21 for Zr-Cu-Ni-Al and 0.4 for La-Co-Al. As the holding time decreases to 100 s, m precipitously reduces to below 0.04 for both samples, which can be more clearly recognized in the inset of Figure 10a. Then, the decrease in m tends to be stable, which are 0.02 and 0.025 in the 15-s holding tests for Zr-Cu-Ni-Al and La-Co-Al, respectively. The holding time effect indicates that strain rate sensitivity could significantly grow from transient creep to steady-state creep. What is more, m is always larger in La-Co-Al than in Zr-Cu-Cu-Ni at all events, while the m gap between the two samples is gradually narrowed as the holding time is shortened, and the values are very close in the 50-s and 15-s holding tests. For long-term holding, the creep flow could turn into a steady-state stage, the deformation behavior of which mainly depends on the intrinsic plastic resistance of the sample. For short-term holding, however, the apparatus error-induced fluctuation plays an important role in the recorded indenter displacement at the transient creep, such as the indenter "overshot" at the beginning of the holding stage, which weakens the creep discrepancy between La-Co-Al and Zr-Cu-Ni-Al. Thus, the variation in the m gap between the two samples in Figure 10a could be qualitatively explained.

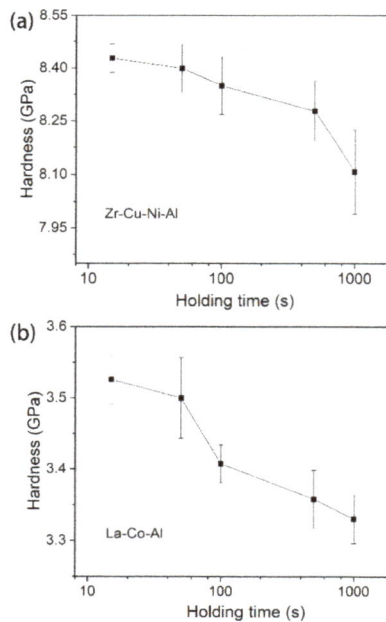

Figure 9. Hardness at the onset of the holding stage as a function of holding time for (**a**) Zr-Cu-Ni-Al and (**b**) La-Co-Al.

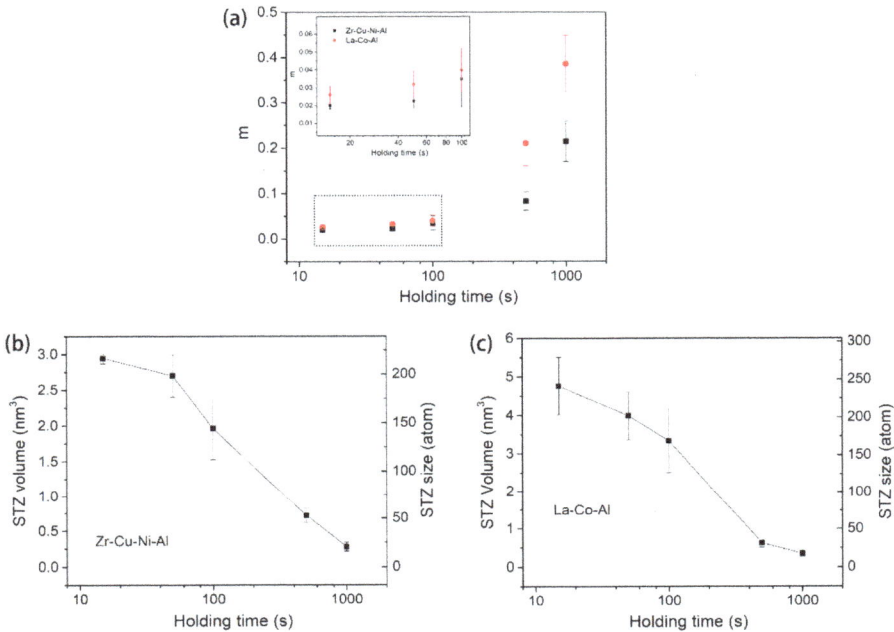

Figure 10. (**a**) strain rate sensitivity as a function of holding time for both samples; STZ size and volume as a function of holding time for (**b**) Zr-Cu-Ni-Al and (**c**) La-Co-Al.

The STZ volume and atoms at each holding event are calculated and summarized in Figure 10b,c. For both samples, STZ sizes shrunk from about 230 atoms to 18 atoms as the holding stage was extended from 15 s to 1000 s. This strong holding time effect on STZ size could be conceivable as being mainly due to the conspicuous enlargement of strain rate sensitivity. Being different from the mechanism of the indentation size effect on STZ size, the holding time effect described here is dominated by the extrinsic method. The different creep flow stages could be the key reason for the large change in STZ size. The apparatus error for short duration, as well as the thermal drift in long-term holding, could also be considerable for the duration-dependent STZ size.

4. Conclusions

In summary, nanoindentation creep measurements were conducted on as-cast Zr-Cu-Ni-Al and La-Co-Al metallic glassy films with a standard Berkovich tip. Based on the strain rate sensitivity of the creep curve, the shear transformation zone (STZ) size was estimated and its correlation with the testing conditions was revealed. The effects of indentation size, loading rate, and holding time on STZ size were systematically studied. Several conclusions could be drawn:

(1) The estimated STZ sizes using the creep method correspond well with the theoretical simulation. Under the same testing conditions, STZ size is detected to be larger in Zr-Cu-Ni-Al than in La-Co-Al, which could be due to the higher glass transition temperature (T_g).

(2) The indentation size effect can be clearly observed in both samples; STZ size increases quickly with increasing pressed depth.

(3) The loading rate effect on STZ size is weak and composition-dependent. STZ size decreases slightly in Zr-Cu-Ni-Al and changes little in La-Co-Al, even though the loading rate is increased by more than an order of magnitude.

(4)   Holding time is particularly important in estimating STZ size. STZ size would artificially decrease from several hundreds of atoms to no more than twenty as the holding stage increases from 15 s to 1000 s.

Author Contributions: T.Z. supervised the research; Y.M. conceived the idea and designed the experiments; Y.S., X.H. and Z.C. interpreted the data and discussed the results; Y.M. wrote the original draft and T.Z. reviewed the final manuscript.

Funding: This research was funded by the National Natural Science Foundation of China (Grant Nos. 11502235 and 11727803) and the Zhejiang Provincial Natural Science Foundation of China (LY18E010006).

Acknowledgments: The authors are also thankful to the staff of Zhejiang University of Technology (ZJUT) for their kind assistance with the experimentation.

Conflicts of Interest: The authors declare no conflict of interest.

References

1.   Klement, W.; Willens, R.; Duwez, P. Non-crystalline structure in solidified gold-silicon alloys. *Nature* **1960**, *187*, 869–870. [CrossRef]

2.   Loffler, J.F. Bulk metallic glasses. *Intermetallics* **2003**, *11*, 529–540. [CrossRef]

3.   Schuh, C.A.; Hufnagel, T.C.; Ramamurty, U. Mechanical behavior of amorphous alloys. *Acta Mater.* **2007**, *55*, 4067–4109. [CrossRef]

4.   Inoue, A.; Shen, B.; Koshiba, H.; Kato, H.; Yavari, A.R. Cobalt-based bulk glassy alloy with ultrahigh strength and soft magnetic properties. *Nat. Mater.* **2003**, *2*, 661–663. [CrossRef] [PubMed]

5.   Leamy, H.J.; Chen, H.S.; Wang, T.T. Plastic flow and fracture of metallic glass. *Metall. Trans.* **1972**, *3*, 699–708. [CrossRef]

6.   Argon, A. Plastic deformation in metallic glasses. *Acta Metal. Mater.* **1979**, *27*, 47–58. [CrossRef]

7.   Mayr, S.G. Activation energy of shear transformation zones: A key for understanding rheology of glasses and liquids. *Phys. Rev. Lett.* **2006**, *97*, 195501. [CrossRef] [PubMed]

8.   Manning, M.L.; Langer, J.S.; Carlson, J.M. Strain localization in a shear transformation zone model for amorphous solids. *Phys. Rev. E* **2007**, *76*, 056106. [CrossRef] [PubMed]

9.   Zink, M.; Samwer, K.; Johnson, W.L.; Mayr, S.G. Plastic deformation of metallic glasses: Size of shear transformation zones from molecular dynamics simulations. *Phys. Rev. B* **2006**, *73*, 172203. [CrossRef]

10. Li, L.; Homer, E.R.; Schuh, C.A. Shear transformation zone dynamics model for metallic glasses incorporating free volume as a state variable. *Acta Mater.* **2013**, *61*, 3347–3359. [CrossRef]

11. Johnson, W.; Samwer, K. A Universal Criterion for Plastic Yielding of Metallic Glasses with a $(T/Tg)^{2/3}$ Temperature Dependence. *Phys. Rev. Lett.* **2005**, *95*, 195501. [CrossRef] [PubMed]

12. Pan, D.; Inoue, A.; Sakurai, T.; Chen, M.W. Experimental characterization of shear transformation zones for plastic flow of bulk metallic glasses. *Proc. Natl. Acad. Sci. USA* **2008**, *105*, 14769–14772. [CrossRef] [PubMed]

13. Kramer, L.; Maier-Kiener, V.; Champion, Y.; Sarac, B.; Pippan, R. Activation volume and energy of bulk metallic glasses determined by nanoindentation. *Mater. Des.* **2018**, *155*, 116–124. [CrossRef]

14. Maier-Kiener, V.; Durst, K. Advanced Nanoindentation Testing for Studying Strain-Rate Sensitivity and Activation Volume. *JOM* **2017**, *69*, 2246–2255. [CrossRef] [PubMed]

15. Pan, D.; Yokoyama, Y.; Fujita, H.; Liu, T.Y.; Kohara, H.S.; Inoue, A.; Chen, M.W. Correlation between structural relaxation and shear transformation zone volume of a bulk metallic glass. *Appl. Phys. Lett.* **2009**, *95*, 141909. [CrossRef]

16. Bhattacharyya, A.; Singh, G.; Prasad, K.E.; Narasimhan, R.; Ramamurty, U. On the strain rate sensitivity of plastic flow in metallic glasses. *Mat. Sci. Eng. A* **2015**, *625*, 245–251. [CrossRef]

17. Choi, I.C.; Yoo, B.G.; Kim, Y.J.; Jang, J.I. Indentation creep revisited. *J. Mater. Res.* **2012**, *27*, 3–11. [CrossRef]

18. Oliver, W.C.; Pharr, G.M. An improved technique for determining hardness and elastic-modulus using load and displacement sensing indentation experiments. *J. Mater. Res.* **1992**, *7*, 1564–1583. [CrossRef]

19. Ginder, R.S.; Nix, W.D.; Pharr, G.M. A simple model for indentation creep. *J. Mech. Phys. Solids* **2018**, *112*, 552–562. [CrossRef]

20. Yoo, B.G.; Kim, Y.J.; Choi, I.C.; Shim, S.T.; Tsui, Y.; Bei, H.B.; Ramamurty, U.; Jang, J.I. Increased time-dependent room temperature plasticity in metallic glass nanopillars and its size-dependency. *Int. J. Plast.* **2012**, *37*, 108–118. [CrossRef]

21. Feng, G.; Ngan, A.H.W. Creep and strain burst in indium and aluminium during nanoindentation. *Scr. Mater.* **2001**, *45*, 971–976. [CrossRef]

22. Afrin, N.A.; Ngan, H.W. Creep of micron-sized Ni₃Al columns. *Scr. Mater.* **2006**, *54*, 7–12. [CrossRef]

23. Chen, J.; Bull, S.J. The investigation of creep of electroplated Sn and Ni–Sn coating on copper at room temperature by nanoindentation. *Surf. Coat. Technol.* **2009**, *203*, 1609–1617. [CrossRef]

24. Zhang, T.H.; Ye, J.H.; Feng, Y.H.; Ma, Y. On the spherical nanoindentation creep of metallic glassy thin films at room temperature. *Mater. Sci. Eng. A* **2017**, *685*, 294–299. [CrossRef]

25. Ma, Y.; Ye, J.H.; Peng, G.J.; Zhang, T.H. Nanoindentation study of size effect on shear transformation zone size in a Ni–Nb metallic glass. *Mater. Sci. Eng. A* **2015**, *627*, 153–160. [CrossRef]

26. Ma, Y.; Peng, G.J.; Jiang, W.F.; Chen, H.; Zhang, T.H. Nanoindentation study on shear transformation zone in a Cu-Zr-Al metallic glassy film with different thickness. *J. Non-Cryst. Solids* **2016**, *442*, 67–72. [CrossRef]

27. Ma, Y.; Peng, G.J.; Feng, Y.H.; Zhang, T.H. Nanoindentation investigation on the creep mechanism in metallic glassy films. *Mater. Sci. Eng. A* **2016**, *651*, 548–555. [CrossRef]

28. Chen, H.; Song, Y.X.; Zhang, T.H.; Ma, Y. Structure relaxation effect on hardness and shear transformation zone volume of a Ni-Nb metallic glassy film. *J. Non-Cryst. Solids* **2018**, *499*, 257–263. [CrossRef]

29. Chen, H.; Zhang, T.H.; Ma, Y. Effect of applied stress on the mechanical properties of a Zr-Cu-Ag-Al bulk metallic glass with two different structure states. *Materials* **2017**, *10*, 711. [CrossRef] [PubMed]

30. Li, H.; Nag, A.H.W. Indentation size effects on the strain rate sensitivity of nanocrystalline Ni–25at. % Al thin films. *Scr. Mater.* **2005**, *52*, 827–831. [CrossRef]

31. Peykov, D.; Martin, E.; Chromic, R.R.; Gauvin, R.; Trudeau, M. Evaluation of strain rate sensitivity by constant load nanoindentation. *J. Mater. Sci.* **2012**, *47*, 7189–7200. [CrossRef]

32. Li, W.H.; Shin, K.; Lee, C.G.; Wei, B.C.; Zhang, T.H.; He, Y.Z. The characterization of creep and time-dependent properties of bulk metallic glasses using nanoindentation. *Matere. Sci. Eng. A* **2008**, *478*, 371–375. [CrossRef]

33. Wang, F.; Li, J.M.; Huang, P.; Wang, W.L.; Lu, T.J.; Xu, K.W. Nanoscale creep deformation in Zr-based metallic glass. *Intermetallics* **2013**, *38*, 156–160. [CrossRef]

34. Yoo, B.G.; Oh, J.H.; Kim, Y.J.; Park, K.W.; Lee, J.C.; Jang, J.I. Nanoindentation analysis of time-dependent deformation in as-cast and annealed Cu–Zr bulk metallic glass. *Intermetallics* **2010**, *18*, 1898–1901. [CrossRef]

35. Ma, Y.; Peng, G.J.; Wen, D.H.; Zhang, T.H. Nanoindentation creep behavior in a CoCrFeCuNi high-entropy alloy film with two different structure states. *Mater. Sci. Eng. A* **2015**, *621*, 111–117. [CrossRef]

36. Ma, Y.; Feng, Y.H.; Debela, T.T.; Zhang, T.H. Nanoindentation study on the creep characteristics of high-entropy alloy films: Fcc versus bcc structures. *Int. J. Refract. Met. H* **2016**, *54*, 395–400. [CrossRef]

37. Chen, J.; Bull, S.J. On the relationship between plastic zone radius and maximum depth during nanoindentation. *Surf. Coat. Technol.* **2006**, *201*, 4289–4293. [CrossRef]

38. Chen, J.; Bull, S.J. On the factors affecting the critical indenter penetration for measurement of coating hardness. *Vacuum* **2009**, *83*, 911–920. [CrossRef]

39. Li, H.; Nag, A.H.W. Size effects of nanoindentation creep. *J. Mater. Res.* **2004**, *19*, 513–522. [CrossRef]

40. Wei, Q.; Cheng, S.; Ramesh, K.T.; Ma, E. Effect of nanocrystalline and ultrafine grain sizes on the strain rate sensitivity and activation volume: Fcc versus bcc metals. *Mater. Sci. Eng. A* **2004**, *381*, 71–79. [CrossRef]

41. Bernal, J.D. Geometry of the structure of monatomic liquid. *Nature* **1960**, *185*, 68–70. [CrossRef]

42. Saha, R.; Nix, W.D. Effects of the substrate on the determination of thin film mechanical properties by nanoindentation. *Acta. Mater.* **2002**, *50*, 23–38. [CrossRef]

43. Chen, J.; Bull, S.J. Relation between the ratio of elastic work to the total work of indentation and the ratio of hardness to Young's modulus for a perfect conical tip. *J. Mater. Res.* **2009**, *24*, 590–598. [CrossRef]

44. Chen, J.; Bull, S.J. A critical examination of the relationship between plastic deformation zone size and Young's modulus to hardness ratio in indentation testing. *J. Mater. Res.* **2006**, *21*, 2617–2627. [CrossRef]

45. Schuh, C.A.; Nieh, T.G. A nanoindentation study of serrated flow in bulk metallic glasses. *Acta Mater.* **2003**, *51*, 87–99. [CrossRef]

46. Ma, Y.; Peng, G.J.; Feng, Y.H.; Zhang, T.H. Nanoindentation investigation on creep behavior of amorphous CuZrAl/nanocrystalline Cu nanolaminates. *J. Non-Cryst. Solids* **2017**, *465*, 8–16. [CrossRef]

47. Yu, H.B.; Wang, W.H.; Bai, H.Y.; Wu, Y.; Chen, M.W. Relating activation of shear transformation zones to β relaxations in metallic glasses. *Phys. Rev. B* **2010**, *81*, 220201. [CrossRef]
48. Cheng, Y.Q.; Ma, E. Atomic-level structure and structure–property relationship in metallic glasses. *Prog. Mater. Sci.* **2011**, *56*, 379–473. [CrossRef]
49. Choi, I.C.; Zhao, Y.; Kim, Y.J.; Yoo, B.G.; Suh, J.Y.; Ramamurty, U.; Jang, J.I. Indentation size effect and shear transformation zone size in a bulk metallic glass in two different structural states. *Acta Mater.* **2012**, *60*, 6862–6868. [CrossRef]
50. Ma, Y.; Peng, G.J.; Debela, T.T.; Zhang, T.H. Nanoindentation study on the characteristic of shear transformation zone volume in metallic glassy films. *Scr. Mater.* **2015**, *108*, 52–55. [CrossRef]
51. Lu, J.; Ravichandran, G.; Johnson, W.L. Deformation behavior of the $Zr_{41.2}Ti_{13.8}Cu_{12.5}Ni_{10}Be_{22.5}$ bulk metallic glass over a wide range of strain-rates and temperatures. *Acta Mater.* **2003**, *51*, 3429–3443. [CrossRef]
52. Choi, I.C.; Zhao, Y.; Yoo, B.G.; Kim, Y.J.; Suh, J.Y.; Ramamurty, U.; Jang, J.I. Estimation of the shear transformation zone size in a bulk metallic glass through statistical analysis of the first pop-in stresses during spherical nanoindentation. *Scr. Mater.* **2012**, *66*, 923–926. [CrossRef]

![micromachines logo] *micromachines*

MDPI

Article

Nanoindentation and TEM to Study the Cavity Fate after Post-Irradiation Annealing of He Implanted EUROFER97 and EU-ODS EUROFER

Marcelo Roldán [1,2,*], Pilar Fernández [1], Joaquín Rams [2], Fernando José Sánchez [1] and Adrián Gómez-Herrero [3]

[1] Division of Fusion Technologies, National Fusion Laboratory, CIEMAT, Avda. Complutense 40, 28040 Madrid, Spain; pilar.fernandez@ciemat.es (P.F.); fernandojose.sanchez@ciemat.es (F.J.S.)
[2] Department of Applied Mathematics, Materials Science and Engineering and Electronic Technology, School of Experimental Sciences and Technology, Rey Juan Carlos University, C/Tulipán s/n, 28933 Móstoles, Spain; joaquin.rams@urjc.es
[3] Centro Nacional de Microscopía Electrónica, Universidad Complutense de Madrid, 28040 Madrid, Spain; adriangh@pdi.ucm.es
* Correspondence: marcelo.roldan@ciemat.es; Tel.: +34-676-824-178

Received: 25 October 2018; Accepted: 23 November 2018; Published: 29 November 2018

check for updates

Abstract: The effect of post-helium irradiation annealing on bubbles and nanoindentation hardness of two reduced activation ferritic martensitic steels for nuclear fusion applications (EUROFER97 and EU-ODS EUROFER) has been studied. Helium-irradiated EUROFER97 and EU-ODS EUROFER were annealed at 450 °C for 100 h in an argon atmosphere. The samples were tested by nanoindentation and studied by transmission electron microscopy extracting some focused ion beam lamellae containing the whole implanted zone ($\approx$50 μm). A substantial increment in nanoindentation hardness was measured in the area with higher helium content, which was larger in the case of EUROFER97 than in EU-ODS EUROFER. In terms of microstructure defects, while EU-ODS EUROFER showed larger helium bubbles, EUROFER97 experienced the formation of a great population density of them, which means that the mechanism that condition the evolution of cavities for these two materials are different and completely dependent on the microstructure.

Keywords: nanoindentation; reduced activation ferritic martensitic (RAFM) steels; helium irradiation; irradiation hardening; nuclear fusion structural materials

---

1. Introduction

The study of the effects of post-irradiation annealing on the evolution of the irradiation defects in which He is involved has a special relevance since its mechanisms and the variables on which those depend are not fully understood, becoming more complicated when the materials investigated are microstructurally complex like the ones studied in this paper. The experimental results referring to nucleation, fate, and consequences of helium irradiation on mechanical properties are especially useful, because they provide information to establish a correlation between experimental results and modelling [1]. In addition, these experiments may also be useful to provide experimental results to validate solubility values and diffusion coefficient of He in its different forms (interstitial atomic He, He-Vacancy clusters (HeV) of different sizes, or even more complex clusters of He and vacancies along with lattice atoms of Fe, HeVFe [2], etc.) which are likewise indispensable when performing modelling work.

In order to validate the theories and models on bubble growth after post-irradiation annealing, it is necessary to examine the effects of some experimental parameters on the evolution of bubbles during annealing by means of experiments under comparable conditions in materials similar to those studied in this research. With the methodology carried out, it has been possible to obtain two structural materials implanted with different levels of He concentration that represent the irradiation expected in a fusion reactor after its useful life (400 to 700 appm) [3,4]. Then, the materials have been annealed at 450 °C, the temperature that has been observed to be critical in the degradation of mechanical properties in materials submitted to irradiation [5]. Numerous authors have indicated at that at 450 °C the greatest volumetric fraction is produced by the formation of cavities generated by irradiation and, therefore, causing the highest degradation of mechanical properties [6–8]. Therefore, in order to understand how temperature affects the evolution of cavities for a given time, an annealing treatment has been applied at 450 °C for 100 h in EUROFER97 and EU-ODS EUROFER steels. He desorption experiments with EUROFER97 [9,10] showed that in a temperature range above 450 °C, the dissociation and diffusion phenomena begin to be critical, so to be able to understand the nucleation and growth of He bubbles, it is necessary to evaluate the mechanisms that govern the diffusion of the He atoms prevailing when the implanted materials are undergoing an annealing process, and to try to find a model to frame that behavior. There are different diffusion mechanisms [11] and its relevance depends on factors such as the nature of the He defect, the annealing temperature, and defects present in the material. Each type of defect involving He (Interstitial He, Substitutional He, or HeV clusters) will present different dissociation and migration energies and, therefore, different easiness to move through the crystalline lattice of the material.

A direct consequence of modifying the microstructure of a material is the change of mechanical properties. It is well known that in general terms, He irradiation produces local hardening [12–14] and a very useful technique to measure this change is nanoindentation, since He irradiation produces a very shallow layer of modified material. This technique has been widely used to measure hardening due to irradiation [8,15], and it also can be used to establish a correlation between microstructural defects and the aforementioned hardening using a model, such as the dispersed barrier hardening one [16,17]. On one hand, cavity density and size are experimental parameters that can be obtained by TEM (Transmission Electron Microscopy). On the other hand, nanoindentation results evidence the effect of voids, cavities, He bubbles, and dislocations produced after He irradiation as they act as barriers for the dislocations generated by the indenter [18]. So, a model combining both would explain the materials hardening mechanisms.

The reduced activation ferritic martensitic (RAFM) steels EUROFER97 and EU-ODS EUROFER have not been studied extensively after implantation and annealing, although some tests have been performed on similar alloys such as F82H [19]. These materials are very important from the point of view of a nuclear fusion reactor, as they are the most promising structural materials as they are able to withstand the extremely harsh conditions which will be produced in the reactor during operation [20,21]. So, this research is a starting point to understand the growth of the complex defects during the annealing treatment at high temperatures of irradiated steels, which may eliminate the pre-nucleation structure and transforms it into a bubble nucleus (or embryo). The subsequent growth of this embryo or of this already formed bubble will be carried out by means of the mechanisms mentioned above if the necessary conditions of time, temperature, and concentration are met: migration and coalescence [22] or Ostwald ripening [23–26].

2. Experimental Procedure

2.1. Materials

The materials investigated in this research were the reduced activation ferritic/martensitic steels EUROFER97 and EU-ODS EUROFER. Both alloys have identical chemical composition (wt. %): 0.11C, 8.7Cr, 1W, 0.10Ta, 0.19V, 0.44Mn, 0.004S, balance Fe. However, EU-ODS EUROFER contains 0.3% of Y_2O_3 particles. On one hand, EU-ODS EUROFER has a ferritic matrix with a large range of grain sizes, showing an average value of 0.98 ± 0.48 µm, although it is possible to find grains as large as 4 µm and others smaller than 0.5 µm. An EBSD study was published elsewhere regarding this matter [27]. Yttria particles with 20 nm size in average were added to the matrix, but their distribution was not completely homogeneous, finding both, small clusters of them and large areas with no particles. On the other hand, EUROFER97 presents a fully martensitic matrix, whose primary austenite grain size is between 6.7 and 11 µm and the average size of its martensite laths is between 0.3 and 0.7 µm. In addition, EUROFER97 has equiaxed morphology, in contrast to EUODS EUROFER with ferritic tangle grains. The steels have been studied in the normalized (980 °C/27 min air cooled) plus tempered (760 °C/90 min air cooled) condition for EUROFER97 (Heat E83698) and normalized (1150 °C/60 min air cooled) plus tempered (750 °C/120 min air cooled) for EU-ODS-EUROFER (Heat HXXX1115), denominated in this paper as the as-received states. A more detailed chemical composition and microstructural characteristics for both materials are given elsewhere [28,29].

2.2. Irradiation and Thermal Treatment

Ion irradiation with He-ions was performed on a 5 MV voltage terminal in a Tandetron accelerator manufactured by High Voltage Engineering Europa (Amersfoort, The Netherlands) at room temperature. A stair-like profile configuration was used for the implantation energy, starting with 15 MeV and ending with 2 MeV, decreasing the beam energy in 1 MeV steps. The ion fluence used for all the energies was 1.67×10^{15} He·cm^{-2}. The resulting He concentration profile was simulated by means of MARLOWE Code [30,31] as shown previously [15,27]. The mean damage produced on the sample was estimated to be around 10^{-2} to 10^{-3} dpa, with an average He injection rate of 0.25–0.3 appm He/s.

During implantation, the temperature of the samples was constantly monitored with a thermographic camera and one thermocouple attached to the sample holder. The maximum temperature measured during implantation was 70 °C.

After irradiation, the specimens were annealed at 450 °C for 100 h in a controlled argon atmosphere.

2.3. Nanoindentation Tests

The equipment used for nanoindentation was a Nano indenter XP manufactured by MTS Systems Corporation (Eden Prairie, MN, USA) equipped with a Berkovich diamond indenter tip. The indentation module used was the one called Quasi-static (QS). The specimens for nanoindentation tests were mechanically polished, finishing with an acidic aluminum oxide suspension in order to achieve a deformation-free surface.

The indentations on the implanted samples of both steels were performed on the transversal section to the irradiation beam, in order to evaluate the change of the hardness values along of implantation depth to get a hardness profile as a function of the He content after the heat treatment.

He load used was 5 mN, which is the load previously used by Roldan et al. [15,27] to measure the hardness variation on He implanted specimens. However, the methodology to obtain results was improved in terms of time-saving, using in this work only one indentation row tilted 10° with respect to the implantation surface. The distance between consecutive indentations was in all tests above 3 times the imprint size. The Oliver & Pharr method was used to determine the mechanical properties from the indentation curves [32,33].

At the beginning of the study, an analysis of the state of the indenter tip is highly recommended. Figure 1 shows an image obtained by optical profilometer in which it is possible to evaluate the curvature radius of the very tip of the indenter, which is critical for the correct evaluation of the indentation data, especially at shallow depths [34,35]. In this case, the quality of the tip was good enough, just showing some dirt, which is easily removed in fused silica during the alignment of the equipment before making the indentations in the specimens.

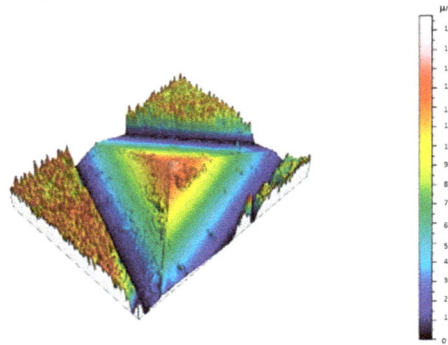

Figure 1. Berkovich indenter tip analyzed by optical profilometer to check the rounding and wear of the very tip.

2.4. Transmission Electron Microscopy

Microstructural investigations on implanted samples of both alloys have been carried out using lamellae extracted by Focus Ion Beam (FIB) in a Zeiss Auriga Compact with a field emission scanning electron microscope (FESEM) of 30 keV. A minimum Ga ion energy of 5 keV was used in the last step of lamella thinning in order to remove as much as possible the Ga damage on the surface [36,37]. Due to the different He concentration along the irradiated volume, it was very important to extract a lamella long enough to allow studying the effect of the annealing with different He concentration, since it is a critical parameter when He bubble fate is studied. Transmission electron microscopy (TEM) investigations were performed with a JEM 2100 HT at 200 keV and JEM 300 F at 300 keV (manufactured by JEOL, Akishima, Japan). In order to calculate a volumetric distribution density of bubbles, it is necessary to determine the thickness of the lamellae, so convergent beam electron diffraction (CBED) was used [38–40].

3. Results

The studies of EUROFER97 and EU-ODS EUROFER after He implantation previously performed showed a maximum increase of hardness of 41% and 21%, respectively, and corresponded to the zone with the highest He concentration [27]. In the following paragraphs the effect of a post-irradiation annealing of 450 °C during 100 h with regard to nanoindentation hardness and irradiation defects will be described in detail.

3.1. Nanoindentation

3.1.1. EUROFER97

The hardness results obtained for EUROFER97 implanted with He from 2 to 15 MeV and consequently annealed at 450 °C for 100 h is represented in Figure 2a in black squares. The dotted red line represents the hardness values for the as implanted condition and the blue dashed line corresponds to the as received state. It is possible to observe that in the annealed state there was a remarkable increase in the hardness values from those of the as received and annealed states, reaching up to 9 GPa, which turns out to be an increase of 157%. There was also a gradual reduction of hardness when the indenting in areas with lower irradiation, reaching the as-received hardness values (~3.5 GPa) at about 25–30 μm from the implantation surface.

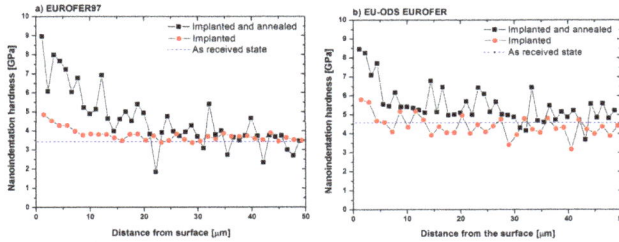

Figure 2. Hardness values vs. distance from the implanted surface of (**a**) EU-ODS EUROFER and (**b**) EUROFER97 irradiated with He from 2 to 15 MeV: Black dotted line represents the values for the annealed state at 450 °C for 100 h. Red dotted are the hardness values for the implanted at RT previous annealing and blue dashed line represents the as received hardness values. All the measurements were performed at 5 mN.

3.1.2. EU-ODS EUROFER

In a similar way to EUROFER97, EU-ODS EUROFER was implanted with He from 2 to 15 MeV and annealed at 450 °C for 100 h. The results from the indentations applied with a load of 5 mN are shown in Figure 2b. In this case, the hardness increments experienced are lower than those observed for EUROFER97, since the as received hardness value of EU-ODS EUROFER was 4.6 GPa, and the maximum hardness value measured on the annealed specimen was 8.49 GPa, which results in an increase of 84.5%. However, beyond 30 μm from the implanted surface, the hardness values that diminish from the surface, reached the same hardness values of the as received state. Again, the effect of the annealing is evident as the slight increase observed after irradiation is lower than that observed after the annealing heat treatment.

3.2. TEM Characterization

TEM investigations were performed on the implanted surface to up to 30 microns in depth in order to evaluate He nucleation and growth due to the heat treatment, and to eventually correlate those observations with the nanoindentation results obtained.

The observation area was divided in 3 zones, each one with a different He content as observed in Figure 3. Zone A, the one with the highest He content, included the peaks from 3 to 5 MeV (716–657 appm He), zone B included the peaks from 6 to 8 MeV (619 to 548 appm He) and the last zone, C, contained the peaks from 9 to 11 (518 to 445 appm He). It was not possible to study the 2 MeV energy peak, since the first 3 to 5 μm resulted damaged in the fabrication process of the lamella.

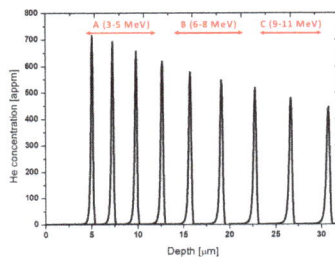

Figure 3. He concentration profiles simulated by MARLOWE indicating the 3 zones in which the lamella was divided for its microstructural characterization by means of TEM.

3.2.1. Cavity Characterization on EUROFER97

The microstructural study by TEM was performed analyzing a lamella following the implantation direction and, in consequence, along the gradual He content reduction. Figure 4 shows a TEM overview of the lamella extracted from EUROFER97 implanted from 2 to 15 MeV and annealed at 450 °C for 100 h.

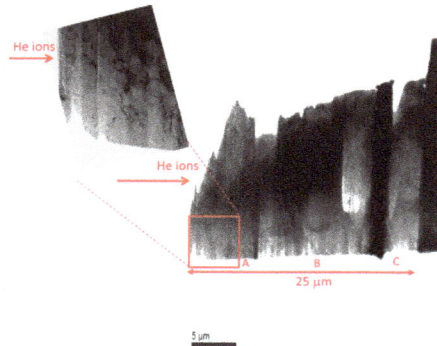

Figure 4. TEM micrograph of the EUROFER97 lamella implanted from 2 to 15 MeV and annealed at 450 °C for 100 h homogeneously thinned along with a detail of the zone A which is the area with the highest He content (716 to 657 appm He).

The cavities were identified using through-focus series method [41] which in addition allows a first approximation of their size. In the in-focus image, most cavities were not clearly detectable, which means that the cavity size was less than 5 nm, approximately as seen in Figure 5 [42,43]. In the so-called zone A, corresponding to the zone with the peaks of 3 to 5 MeV with a He content of 716 to 675 appm He, the steel matrix of EUROFER97 was completely full of small cavities, which were randomly distributed.

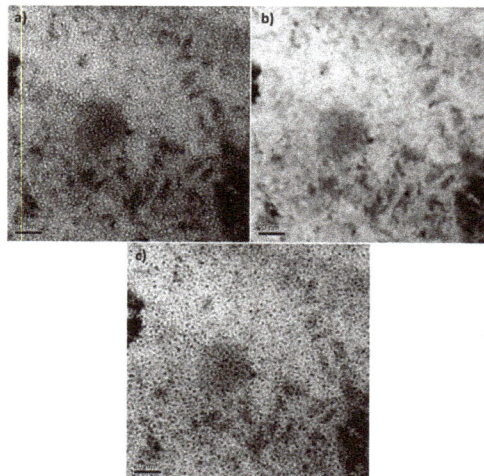

Figure 5. TEM focus sequence of the zone A of EUROFER97 annealed at 450 °C for 100 h and a He content from 716 to 657 appm. The focus variation was about ±1 μm: (**a**) under focused, (**b**) in-focus, and (**c**) over focused.

The vast majority of the cavities exhibited a diameter between 1 and 2 nm, and, only in some isolated cases was the diameter slightly larger, but never reaching 3 nm. Taking into account a lamella thickness of around 60 nm, as determined by CBED [38,39], the population density was 9.65×10^{23} m^{-3}.

The study of cavity distribution did not reveal any preferential nucleation at grain/subgrain boundaries, nor at precipitate–matrix interfaces.

As mentioned at the beginning of this section, if the analysis moves forward following the implantation direction, the implanted He amount decreases gradually up to reach the so-called zone B with a He concentration between 619 to 518 appm. Figure 6 shows cavities found in this zone whose sizes are quite similar to the ones detected at the previous zone, 1 to 2 nm. The cavity distribution was still random and no preferential nucleation was detected. In this zone, the population density was 2.25×10^{23} m^{-3}, so it was slightly lower than the one measured in zone A, although the cavity average size was quite similar.

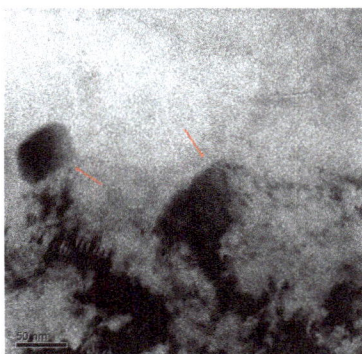

Figure 6. TEM micrograph of EUROFER97 steel annealed at 450 °C for 100 h showing the distribution of bubbles within the matrix in zone B of intermediate He concentration (619 to 548 appm). The arrows indicate bubble nucleation at the $M_{23}C_6$ matrix interface.

In the mentioned figure, the presence of two precipitates type $M_{23}C_6$ with some cavities attached to their matrix-precipitate (indicated with red arrows in the same figure) can be also observed, and in their surroundings. Due to the lack of zones with no cavities around the precipitates (known as depletion zone), it is not possible to confirm completely that those microstructural features were acting as defect sinks before the annealing. As it has been experimentally observed by TEM [44–46], when a microstructural feature acts as a sink, it attracts a HeV cluster flux towards it, creating an area with no defect clusters around the sink. This zone would have certain length that would depend on the vacancy (or HeV cluster) absorption rate which in turn is proportional to the precipitate diffusion and the difference between the defects of the matrix and of the sink surface [47,48].

Finally, to complete the whole TEM characterization of the EUROFER97 lamella, zone C, with He content between 518 and 445 appm He, was analyzed. In Figure 7, it is possible to see in the lower right corner, a black area that was too thick for the electrons to get through. It corresponds with the lamella edge that was welded onto the TEM grid. Unlike A and B areas described previously, groups of very small cavities randomly distributed across the matrix were observed. Some of those clusters have been indicated with arrows and red ovals. The cavity diameters were not larger than 2 nm and the population density was much lower than the one calculated for the other areas, 1.63×10^{23} m^{-3}. In contrast to zones A and B, some free from cavities areas were detected within this zone. It is difficult to determine exactly where these areas were located regarding the direction of implantation, this fact may suggest that in spite of performing an annealing treatment at 450 °C for 100 h, either the time or temperature were not high enough to induce cavities to diffuse in such a way that they would spread all across the irradiation surface, enhancing nucleation or even the growth of already formed ones, in order to obtain a more homogeneous distribution.

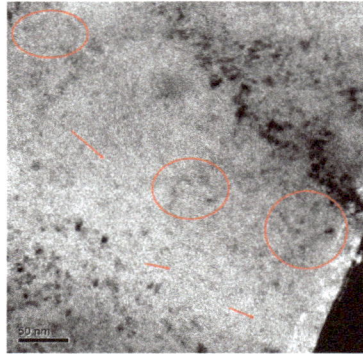

Figure 7. TEM micrograph of EUROFER97 steel annealed at 450 °C for 100 h showing the distribution of bubbles within the matrix in zone C at the lowest He concentration (518 to 445 appm He). Red arrows and ovals highlight some cavities.

3.2.2. Cavity Characterization on EU-ODS EUROFER

Figure 8 shows an overview of the EU-ODS EUROFER lamella extracted from the implanted and annealed bulk specimen. As can be observed, the lamella was fabricated with some separators highlighted with blue arrows in the figure, with the aim of helping with the identification in TEM of the different areas with different He content, as it is required to correlate microstructure with penetration depth and with He content. The red arrow indicates the ion implantation direction and the zones with different ion implantation dose are marked as A, B, and C.

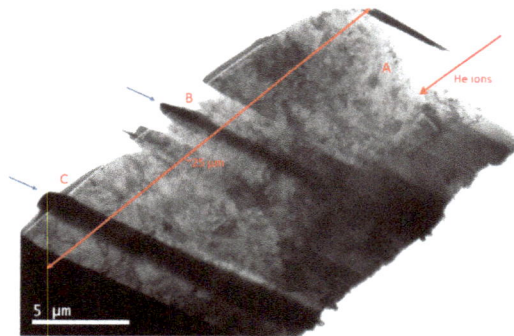

Figure 8. TEM micrograph overview of the lamella obtained from the EU-ODS EUROFER steel annealed at 450 °C for 100 h evenly thinned. The red arrow indicates the implantation surface. A, B, and C indicate areas with different He content and blue arrows show separators.

An exhaustive TEM study was performed in the zone A and in it, some bands of cavities are visible (Figure 9), which may match with a former Bragg's peak. This area is the closest one to the implantation surface and its width is around 7–8 μm. In addition, in this zone, many cavities randomly distributed across the matrix with a very heterogeneous diameter were detected.

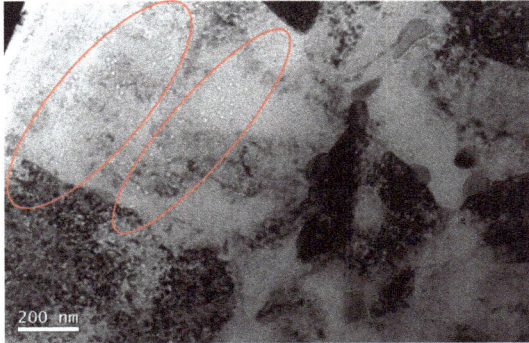

Figure 9. A representative TEM Micrograph of zone A (716 to 657 appm He) of EU-ODS EUROFER steel annealed at 450 °C for 100 h. Highlighted in red the areas of higher cavity density.

EU-ODS EUROFER present Y_2O_3 particles dispersed in the matrix, and due their size and contrast on focus condition in TEM they could be confused with the largest cavities. However, performing the through-focus series, the aforementioned particles' contrast does not change and cannot be considered in the statistics for cavities.

The size of the cavities was measured, although those with diameters lower than 2 nm were not considered due to the difficulty to measure them accurately (the error may be higher than 15%) and because the volumetric fraction is low in comparison with that of larger cavities [49]. Figure 10 shows the coexistence of cavities of different diameters randomly distributed within the ferritic matrix. In addition, there were not clear indications of cavities at grain boundaries or yttria particles.

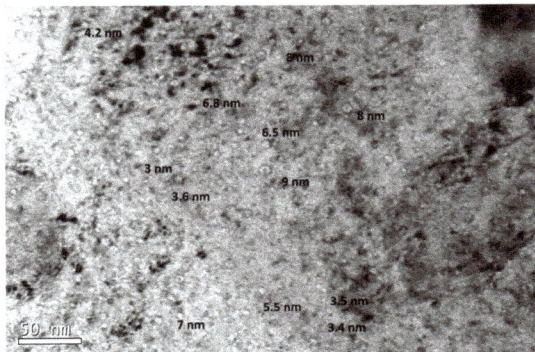

Figure 10. Under-focused TEM micrograph from zone A (716–657 appm He) of EU-ODS EUROFER annealed at 450 °C for 100, including the diameters of some representative cavities.

The size distribution histogram is plotted in Figure 11. The average size was 4.7 ± 1.2 nm, with cavities even larger than 10 nm, in contrast to EUROFER97 zone A where the average size was 2 nm.

Cavities in zone A of EU-ODS EUROFER

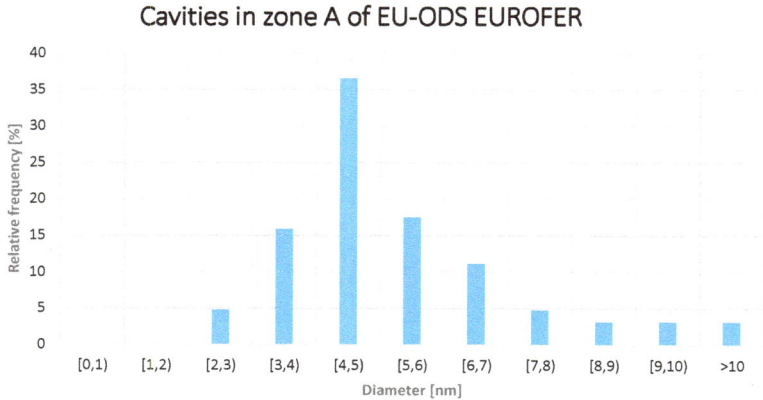

Figure 11. Cavity diameter distribution within the zone A of EU-ODS EUROFER (657–716 appm He) and annealed at 450 °C for 100 h.

As in the case for EUROFER97, assuming that all the cavities are sphere-like, population density was determined considering only the cavities larger or equal to 2 nm. The calculated value was 1.84×10^{22} m^{-3} that corresponds to a volumetric fraction of 0.17%.

The next area of interest, zone B in Figure 8 and with a He content between 619 and 548 appm, is characterized by having a more homogeneous cavity size distribution than the one measured in A, with a Gaussian-like size distribution (Figure 12). Most of the detected cavities presented a diameter between 3 and 4 nm (>35%) and no one with a diameter larger than 7 nm (in contrast to zone A where cavities larger than 10 nm were observed).

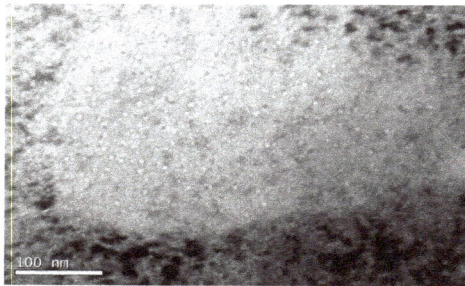

Figure 12. Cavities detected within the zone B of EU-ODS EUROFER (548–619 appm He) annealed at 450 °C for 100 h.

Although the cavity distribution seems to be random, within zone B, cavities attached to the Y_2O_3–matrix interface and grain boundaries were observed (Figure 13). Those observations may be attributable to the random nucleation process itself that may take place before the annealing and the low mobility during the annealing (high migration energy [50]) of large vacancy-helium clusters. Therefore, it is not possible to assert if the addition of yttria particles under these experimental conditions enhance the defect suppression.

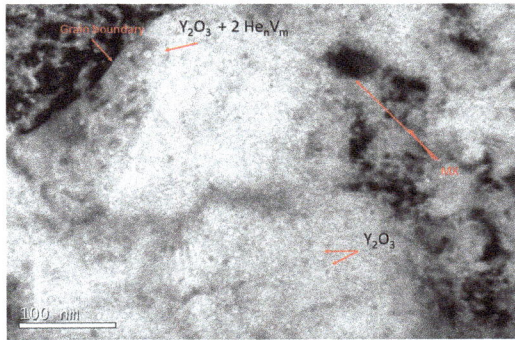

Figure 13. TEM micrograph belonging to zone B (548–619 appm He) of EU-ODS EUROFER steel annealed at 450 °C for 100 h. A grain boundary and two MX type precipitates are highlighted with no clear indication of cavities in their surrounds. In addition, two yttria particles with no cavities along with another one with two cavities attached are also pointed out.

After an exhaustive TEM analysis of different areas belonging to zone B, the distribution density of cavity sizes was calculated following the same procedure than for the zone A, obtaining a value of 3.65×10^{21} m^{-3}, average size of 3.2 ± 1 nm and a volumetric fraction of 0.12%. In Figure 14 the size distribution of the cavities in zone B is shown.

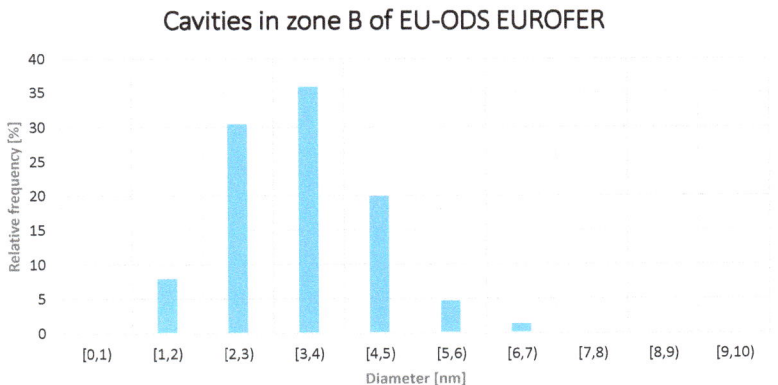

Figure 14. Cavity diameter distribution within the zone B of EU-ODS EUROFER (548 to 619) and annealed at 450 °C for 100 h.

Finally, in zone C with helium content between 445 to 518 appm, the detection of cavities was more complicated. One of the possible reasons added to the lower implanted helium concentration is that the lamella was slightly thicker in this region than in the previous zones (~70 nm on average). This peculiarity is very common and limits the capability of detecting cavities with diameters smaller than 4 nm.

In zone C, the decrease of cavity density was clearly observed (Figure 15a,b). In general terms, EU-ODS EUROFER presented very similar cavities randomly distributed in the matrix with a size ~2 nm. However, a few larger cavities up to 5 nm were detected as well (highlighted in red in Figure 15a).

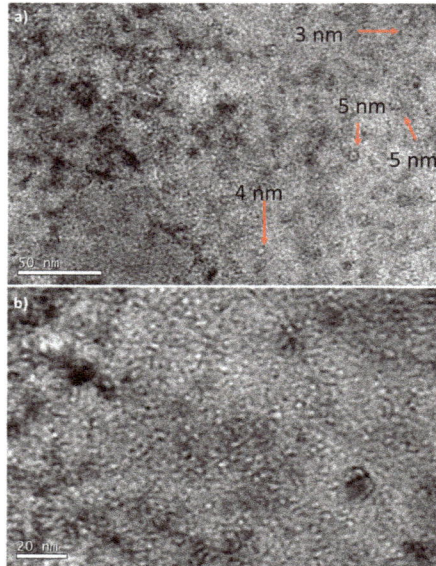

Figure 15. TEM micrograph representative of zone C (518 to 445 appm He) of EU-ODS EUROFER steel annealed at 450 °C for 100 h showing (**a**) very few medium size cavities and (**b**) the overall distribution of small cavities.

Unlike zones A and B, where cavity nucleation was not observed at the grain boundaries (Figure 16), in zone C, some boundaries containing aligned cavities were clearly present. In the right area of Figure 16, two grain boundaries have been highlighted with red ovals, in which a possible alignment of cavities can be observed; they are very similar to those found by Luo et al. [51]. It is difficult to confirm it unequivocally, but as mentioned above, it has been observed by several authors that when grain boundaries act as sinks and trap cavities, an area without cavities appears around these boundaries [48]. In this case, not only there is an area around the grain boundaries free of cavities, but at the right of the red line there appears to be none of relevant size, suggesting that they may be smaller in size, again undetectable by TEM. The value for the distribution density was $9.65 \times 10^{22} \text{ m}^{-3}$ and the volumetric fraction was 0.002%.

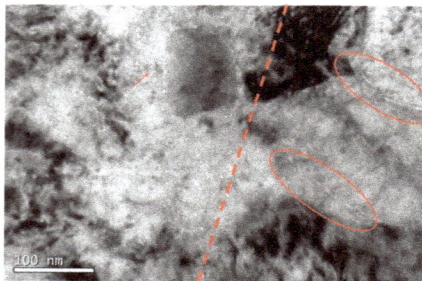

Figure 16. TEM micrograph from zone C of EU-ODS EUROFER steel annealed at 450 °C for 100 h where a possible end of a Bragg peak is indicated as well as different cavity groupings.

Comparing the results for the three zones it is possible to extract that the lower the helium content, the lower the cavity size, with the most typical cavity size being 4 to 5 nm in zone A, 3 to 4 nm in zone B, and 2 nm in zone C. On the other hand, the maximum sizes observed also have changed, decreasing from 10 nm or larger in zone A to a maximum of 6–7 nm in zone B, whereas in zone C, very few cavities were much larger than 2 nm, however, the maximum size detected was 5 nm.

4. Discussion

Regarding nanoindentation results, it has been determined that EUROFER97 and EU-ODS EUROFER steels after irradiation and subsequent annealing at 450 °C experienced a significant increase of hardness. The maximum values are practically identical in both materials, between 8.5 and 9 GPa (Figure 2) measured in the so-called zone A with a high He content. In EU-ODS EUROFER steel, hardness values progressively increase to almost 30 μm depth (~450 appm He), while in EUROFER97 steel, hardness values increase up to a distance of approximately 20 to 30 μm, which is the area of maximum He content (between 400 and 500 appm He). That would correspond in the case of EU-ODS EUROFER steel with the entire lamella depth (zones A, B, and C) and in the case of EUROFER97, the greatest increase occurs mainly in zone A, although also in zone B and to a lesser extent in zone C. The aforementioned increase in hardness in both materials occurs because there are elements that acted as obstacles preventing the sliding of dislocations generated by the indentation. These obstacles are of a very diverse nature: existing dislocations in the material, yttrium oxides (in the case of the EU-ODS EUROFER), precipitates, high quantity of grain edges (small grains), and, of course, the defects produced by irradiation that have evolved due to the annealing treatment, which in this case are HeV clusters, becoming eventually He bubbles (or cavities). When taking into account only the effect of He in the increase of hardness, it is necessary to analyze both the size and the distribution of cavities as possible obstacles for dislocations produced by the indentation volume.

The variation of hardness values of the EU-ODS EUROFER steel is proportional to the size and density of the cavities found in the different areas analyzed. In zone A, the material presents the largest cavity mean size and the widest distribution density, which causes the observed hardness increase. Comparing the results obtained before and after annealing, EU-ODS EUROFER clear change of the irradiations defects observed, since after annealing it experienced a growth of the cavities at the expense of a decrease in distribution density, as is clearly seen in Figure 17a,b.

However, it has been observed that in zone C with a He content between 479 and 445 appm He, EU-ODS EUROFER steels presented some cavities after annealing, not being detected in the window of low implanted He content at room temperature of both materials [27]. Due to the large number of defects (in the form of vacancies and interstitial) present on the EU-ODS EUROFER steel, the thermal diffusion of these defects takes on a special relevance since they move while interacting with the HeV clusters generated by the implantation due to the annealing treatment applied.

In the case of the EUROFER97, a very high increase in zone A (high He content), which has a high density of cavities, was experienced. Figure 17c,d compares the cavities detected in areas with a He content of ~700 appm He in EUROFER97 steel before and after annealing. In this image it is possible to see how the size of the cavities is similar (1 to 2 nm), but the distribution density of these defects from a qualitative point of view seems much higher after the thermal treatment. It should be noted that the annealed steel image was acquired at higher magnifications and, therefore, the cavities appear larger, yet taking a random area in both micrographs results in higher distribution density in the annealed sample.

These hardness values are gradually decreasing, within zones B and C, until they reach values similar to those implanted at room temperature. It should be noted that in the area with the lowest He content of the samples implanted at room temperature without heat treatment no cavities were detected [27], which was the same observation extracted from EUROFER97. However, in the annealed sample there is a substantial increase in population as can be seen in Figure 17d. These results may indicate that due to the effect of temperature there was an increase in nucleation of possible bubble embryos that were detectable by TEM, the mobility of clusters depends completely on their size (the smaller ones present a higher mobility).

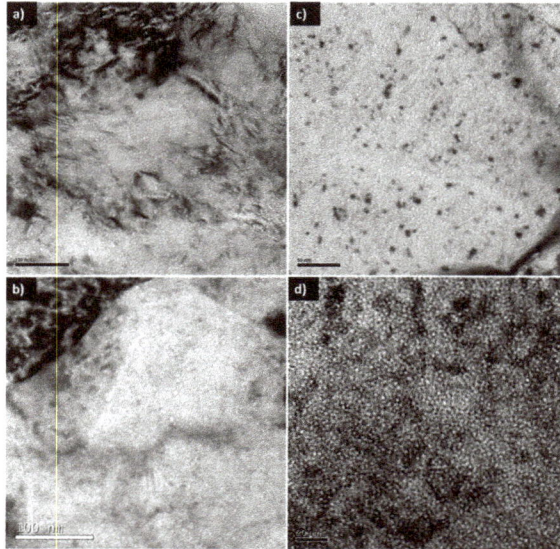

Figure 17. TEM micrographs of the microstructure of EU-ODS EUROFER (**left**) and EUROFER97 (**right**) steel with a He content of ~700 appm He (**a**) and (**c**) implanted at room temperature with stair-like profile from 2 to 15 MeV and (**b**) and (**d**) implanted at room temperature with stair-like profile from 2 to 15 MeV and annealed at 450 °C for 100 h.

Other critical factors are the number of defects in the matrix as well as the microstructural characteristics presented in the material (grain boundaries, precipitates interfaces ...). In the case of materials subjected to neutron irradiation [52,53] or another species that generates cascades of displacements, the formation of vacancies is favored by to the formation of Frenkel pairs by irradiation. This leads to a supersaturation of vacancies, since the self-interstitial atoms are very mobile and can form dislocation loops and be trapped by sinks, which will have different effects on the microstructure of the material, and therefore on the mechanical properties. This supersaturation of vacancies increases the diffusivity of He by providing pathways of diffusion [52,54]. In the case of post-irradiation annealing, there was no increase in vacancies due to new cascades, so the parameters that control the evolution of the bubbles are the vacancies already produced by the implantation of He prior to annealing in combination with thermal vacancies, the temperature and thermal treatment duration, He concentration, and, of course, the type of material and its microstructural characteristics (phases, composition, own defects, concentration of vacancies...). With respect to the results mentioned above, it is possible to explain the very different behavior between EU-ODS EUROFER steel and EUROFER97. The first one presents a much higher concentration of vacancies, which, although not measured experimentally, are assumed to be present due to the manufacturing route: powder metallurgy plus the addition of yttria particles. This material has experienced a considerable increase in the size of the cavities in the area with the highest He content (zone A) with respect to the sample of the same concentration but implanted at room temperature [15,27]. In addition, during annealing, the size of the cavities in zones B and C decreases, as well as their distribution density. The chosen temperature activated the diffusion of vacancy clusters of both the material itself and HeV clusters produced by irradiation and favored the reabsorption of these in the already formed bubbles. As the concentration decreases, the amount of nucleated bubbles prior to annealing decreases, and therefore the surface pressure is lower, causing this mechanism to slow down. On the other hand, in EUROFER97, it has been observed that the distribution density increased with respect to implantation at room temperature; however, the average radius remains constant, between 1 and 2 nm. This fact seems to indicate that the decrease in vacancies compared to the EU-ODS EUROFER, implies a decrease in the diffusion process even though the temperature was 450 °C.

Representing cavity size and population density measured in every zone of the implanted depth, it can be observed that EUROFER97 (Figure 18) does not experience cavity growth: in the three zones the diameter is 1 to 2 nm but there is an enormous increase of population density. There was possibly a growth of small bubble embryos (very small HeV clusters).

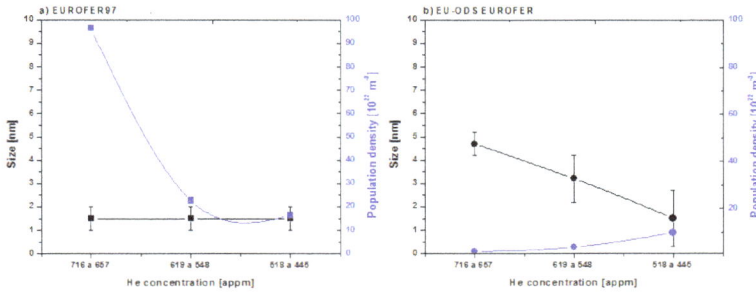

Figure 18. Cavity size and population density measured along the implantation depth profile for (**a**) EUROFER97 and (**b**) EU-ODS EUROFER.

After the exhaustive analysis, we tried to match the experimental observations to either of the two theoretical model proposed to explain the evolution of He bubbles after post-irradiation annealing: Ostwald ripening [23–26] and the theory of migration and coalescence (MC) [22].

The Ostwald ripening model, first described in 1896 [26], explains a very common phenomenon occurring in solid or liquid solutions in which homogeneous structures (such as crystals or sol particles) change size over time. This concept has been applied to the cavities formed by HeV clusters, as follows: it is two relatively close cavities of two different sizes formed by an agglomeration of He atoms and vacancies. These structures are energetically stable except for the He atoms that are located on the surface, which are more energetically unstable than those ones inside. As time goes by, these atoms will dissociate from the smaller cavity, causing a decrease in size, having a lower surface energy than the large cavity and will be introduced into the matrix, increasing the number of atoms in solution. When the matrix is oversaturated with He atoms from the cavity that is shrinking, they will be redeposited into the larger cavity around them, causing it to grow. In addition to the dissociation and reabsorption of He atoms, Ostwald ripening also requires the dissociation and reabsorption of vacancies. So, this process which is subject to the dissociation of two species (He atoms and vacancies) will depend on which of the two dissociation energies is larger [55].

On the other hand, the MC theory attributes the growth of the bubbles to the merge of bubbles that, following a diffusion path through the matrix, finally join forming a larger cavity. Ono et al. [56] studied Fe and Fe9Cr samples implanted with He at 10 keV with different fluences and temperatures and applied afterwards an annealing treatment to the samples in steps of the same time from 400 to 1000 °C. In the mentioned paper, it is possible to see the path followed by a mobile cavity when an annealing treatment at 750 °C is applied to a pure Fe sample irradiated with He at 300 °C with a fluence of 6×10^{19} ions/cm^2. This work provides evidence that bubble mobility is Brownian (or random) and thermally activated, as well as experimental evidence that the bubble mobility in Fe9Cr is lower than in pure Fe. This is in agreement with the aforementioned on the importance of alloying elements in terms of cavity mobility. EUROFER97 and EU-ODS EUROFER are not model alloys with a simple microstructure where the cavities can diffuse through the matrix easily, since there are many barriers that prevent the movement. However, in short range order, cavities close to one another may move and coalescence in a more pronounced way when the density of small cavities is as high as that detected in this paper. These observations fit with the results obtained in modelling works regarding the relation between complexity defect and mobility [2,57].

In EU-ODS EUROFER (Figure 18b), the annealing temperature seems to activate the diffusion of vacancies (both thermal and the ones produced by irradiation) which may favor the reabsorption in already formed bubbles. As He content decreases, the number of cavities already formed also diminishes and therefore the internal pressure as well as the number of atoms in solution and vacancies decrease, slowing down the process. These observations may fit Ostwald ripening. Even so, what can be assured is that the vacancies together with the own defects of the EU-ODS EUROFER steel offer preferential sites to form HeV clusters. In addition, the fact that they do not migrate towards grain and/or yttria edges suggests that this is a very local evolution favored by an average temperature. Even so, what can be assured is that the vacancies together with the own defects of the EU-ODS EUROFER steel offer preferential sites to form HeV clusters. There are signs of a growth of cluster embryos, as there is an increase in the density of cavities in the EUROFER97 compared to those implanted at room temperature, but this evolution does not follow either the MC model or the Ostwald maturation model. In the case of the EU-ODS EUROFER, it does appear that the size of the cavities increases with concentration after annealing, with a decrease in density, apparently indicating that Ostwald maturation is favored by a high number of vacancies, according to Trinkaus [52].

Temperature affects the diffusion of defects as observed by modeling. Many authors [2,57] have observed that the mobility increases with temperature, and it is very dependent on the complexity of the defect. Some observations have experimentally demonstrated [48,58–60] that the most mobile species are the He interstitial atoms rather than vacancies, even at room temperature. They concluded that alloying with up to an amount in the order of ppm may affect the distribution and size of cavities, thus modifying the distance of the zone around the grain edges without bubbles, which is smaller in purer materials than in commercial ones. Comparing these results with those obtained with EUROFER97 and EU-ODS EUROFER, which present a much more complex microstructure in terms of secondary phases and defects, it has been observed that the alloy elements are not affected as much as the model alloys since both materials present the same atomic concentration of the chemical elements. Therefore, the most notable difference is due to the different microstructures and the manufacturing process.

By increasing the concentration of He in the material during implantation, there is an increase of the internal pressure. This could explain the different behavior of the bubbles after post-irradiation annealing as there were different He concentration ranges and therefore different surface pressures. In addition, it is known that the surface is a source of vacancies which are available to relax the internal pressure of bubbles close to the surface [55]. Consequently, a slow growth can be attributed to bubbles with high internal pressure, which can increase in size through migration and coalescence. On the other hand, bubbles that grow rapidly can present Ostwald maturation, which seems to present its maximum value when the pressure is relaxed by the vacancies caused by cascades of displacements produced by irradiation or taken from the surface [55]. Stoller et al. [61] observed that as the annealing temperature increased, the size of the bubbles increased and their density decreased instead. They also detected black dots in the matrix that could be mostly small interstitial clusters [62]. Subsequently, Golubov et al. [63] modeled Stoller's experiment (cited above as [61]) using the theory of bubble migration and coalescence. Without assuming that they remain in mechanical equilibrium during annealing, they concluded that this assumption was quite realistic at high temperatures. Therefore, the bubbles remain over-pressurized during annealing, and this high pressure suppresses surface diffusion and, consequently, growth. On the other hand, by increasing the temperature, the mobility of the vacancy clusters increases, favoring their growth at the same time as their distribution density decreases, as observed experimentally in samples annealed from 700 to 900 °C [61].

On the other hand, having observed zones without cavities, as well as grain boundaries, precipitated interfaces or in the case of the EU-ODS EUROFER, yttria oxides, free of cavities, suggests that reabsorption or growth to be visible in the TEM of new cavities is a short-range, local process, unlike a dislocation network that would be found within the so-called long-range defects. The migration of V-type clusters is carried out by jumps towards the nearest vacancy neighbor, passing through an intermediate metastable state, with its migration energy depending on size as it has been shown above, until reaching V5, when they are considered practically immobile [64]. It is a fact that He implantation damages EU-ODS EUROFER to a lesser extent, since the increase in hardness (which is a direct consequence of irradiation damage) is lower. However, unlike some modelling works [65,66] demonstrating the radiation resistance is only due to yttrium oxide dispersed though the matrix, it is hard to prove since there are other factors such as dislocation density, secondary phases, or inherent vacancies.

5. Conclusions

The most important conclusions extracted from this work are the following:

- The annealing treatment at 450 °C for 100 h has led to an increase in hardness values of 157% for EUROFER97 steel and of 84% for EU-ODS EUROFER with respect to the as-received condition when a load of 5 mN is applied with a Berkovich tip.
- It was experimentally demonstrated that for faster tests, a row matrix is valid to analyze the surface transverse to the implantation, as long as it covers the entire implanted surface and no indentations are duplicated.
- Experimental observations by TEM indicate that EUROFER97 steel experienced an increase in population density of cavities as a function of He concentration. The values of distribution density have been quantified, assuming that most cavity sizes were between 1 and 2 nm. The estimate of the calculated distribution density was 9.6×10^{23} m^{-3} in zone A, 3.25×10^{23} m^{-3} in zone B, and 1.63×10^{23} m^{-3} in zone C. These values suggest that the population density is directly proportional to the concentration of He implanted after the annealing heat treatment at 450 °C for 100 h.
- The EU-ODS EUROFER steel, on the other hand, shows a notable increase in the size of the cavities, which decreases depending on the concentration of He implanted. It should be borne in mind that only cavities larger than 2 nm were taken into consideration; those ones with a smaller diameter were not taken into account due to their low influence on the swelling phenomenon (or volumetric fraction). In zone A, the average size was 4.7 nm with a distribution density of 1.846×10^{22} m^{-3} and a swelling of 0.17%. In zone B, the mean diameter was 3.2 nm, its population density was 3.656×10^{21} m^{-3}, and swelling of 0.12% was calculated, and, finally, in zone C, the average size of the cavities was between 1 and 2 in almost all cases, with some cavities of 4 and even 5 nm. Therefore, its distribution density was calculated analogously to EUROFER97 (9.656×10^{22} m^{-3}) and its volumetric fraction was almost negligible (0.002%). The effect of the inherent vacancies in the EU-ODS EUROFER steel seems to play a very important role, as it is possible that the annealing temperature chosen favors the mobility of these defects, thus enhancing the creation and growth of He cavities. It has also been observed that their size depends strongly on the concentration of He implanted.
- It is not possible to conclude which mechanism governs the nucleation and growth of cavities after a process of annealing at 450 °C for 100 h in both materials when only considering the maturation of the Ostwald and migration and coalescence (MC) models. In order to do so, further experiments would be required to study other annealing times and temperatures and different He concentrations in order to obtain a more complete spectrum of cavity evolution or, if not, to establish a model that fits better than those mentioned in the paper.

Author Contributions: Conceptualization, M.R., P.F. and J.R.; methodology, M.R., F.J.S. and A.G.-H.; formal analysis, M.R., P.F. and J.R.; investigation, M.R. and A.G.-H.; resources, P.F.; data curation, M.R.; writing—original draft preparation, P.F., F.J.S. and M.R.; writing—review and editing, P.F., F.J.S. and M.R.; supervision, P.F. and J.R.; project administration, P.F.; funding acquisition, P.F.

Funding: This work has been supported by Ministerio de Ciencia, Innovación y Universidades Projects: ENE2015-70300-C3-1-R and MAT2012-384407-C03-01, TechnoFusion Project (S2013/MAE-2745) of the CAM (Comunidad Autónoma Madrid) and partially by the European Communities within the European Fusion Technology Programme 2014–2018 under agreement No 633053. "The views and opinions expressed herein do not necessarily reflect those of the European Commission".

Acknowledgments: The authors want also to thank to the National Centre from Electronic Microscopy stuff and all the researchers and technician from CIEMAT. Rey Juan Carlos University and CMAM (Centro de Microanálisis de Materiales) are also acknowledge for helping.

Conflicts of Interest: The authors declare no conflict of interest.

References

1. Schäublin, R.; Chiu, Y.L. Effect of helium on irradiation-induced hardening of iron: A simulation point of view. *J. Nucl. Mater.* **2007**, *362*, 152–160. [CrossRef]
2. Terentyev, D.; Juslin, N.; Nordlund, K.; Sandberg, N. Fast three dimensional migration of He clusters in bcc Fe and Fe–Cr alloys. *J. Appl. Phys.* **2009**, *105*, 103509. [CrossRef]
3. Gilbert, M.R.; Dudarev, S.L.; Zheng, S.; Packer, L.W.; Sublet, J.-C. An integrated model for materials in a fusion power plant: Transmutation, gas production, and helium embrittlement under neutron irradiation. *Nucl. Fusion* **2012**, *52*, 083019. [CrossRef]
4. Knaster, J.; Moeslang, A.; Muroga, T. Materials research for fusion. *Nat. Phys.* **2016**, 424–434. [CrossRef]
5. Aiello, G.; Aktaa, J.; Cismondi, F.; Rampal, G.; Salavy, J.F.; Tavassoli, F. Assessment of design limits and criteria requirements for Eurofer structures in TBM components. *J. Nucl. Mater.* **2011**, *414*, 53–68. [CrossRef]
6. Klimenkov, M.; Möslang, A.; Materna-Morris, E. Helium influence on the microstructure and swelling of 9%Cr ferritic steel after neutron irradiation to 16.3dpa. *J. Nucl. Mater.* **2014**, *453*, 54–59. [CrossRef]
7. Wakai, E.; Sawai, T.; Naito, A.; Jitsukawa, S. Microstructural development and swelling behaviour of F82H steel irradiated by dual ion beams. *J. Electron Microsc.* **2002**, *51*, S239–S243. [CrossRef]
8. Fave, L.; Pouchon, M.A.; Döbeli, M.; Schulte-Borchers, M.; Kimura, A. Helium ion irradiation induced swelling and hardening in commercial and experimental ODS steels. *J. Nucl. Mater.* **2014**, *445*, 235–240. [CrossRef]
9. Carvalho, I.; Schut, H.; Fedorov, A.; Luzginova, N.; Desgardin, P.; Sietsma, J. Helium implanted RAFM steels studied by positron beam Doppler Broadening and Thermal Desorption Spectroscopy. *J. Phys. Conf. Ser.* **2013**, *443*, 012034. [CrossRef]
10. Carvalho, I.; Schut, H.; Fedorov, A.; Luzginova, N.; Sietsma, J. Characterization of helium ion implanted reduced activation ferritic/martensitic steel with positron annihilation and helium thermal desorption methods. *J. Nucl. Mater.* **2013**, *442*, S48–S51. [CrossRef]
11. Trinkaus, H.; Singh, B.N. Helium accumulation in metals during irradiation—Where do we stand? *J. Nucl. Mater.* **2003**, *323*, 229–242. [CrossRef]
12. Takayama, Y.; Kasada, R.; Yabuuchi, K.; Kimura, A.; Hamaguchi, D.; Ando, M.; Tanigawa, H. Evaluation of Irradiation Hardening of Fe-Ion Irradiated F82H by Nano-Indentation Techniques. *Mater. Sci. Forum* **2010**, *654–656*, 4.
13. Wakai, E.; Jitsukawa, S.; Tomita, H.; Furuya, K.; Sato, M.; Oka, K.; Tanaka, T.; Takada, F.; Yamamoto, T.; Kato, Y.; et al. Radiation hardening and -embrittlement due to He production in F82H steel irradiated at 250 °C in JMTR. *J. Nucl. Mater.* **2005**, *343*, 285–296. [CrossRef]
14. Liu, C.; Hashimoto, N.; Ohnuki, S.; Ando, M.; Shiba, K. Dependence of Dose and He on Irradiation-Hardening of Fe-Ion Irradiated Fe8Cr Model Alloy. *Mater. Trans.* **2013**, *54*, 96–101. [CrossRef]
15. Roldán, M.; Fernández, P.; Rams, J.; Jiménez-Rey, D.; Ortiz, C.J.; Vila, R. Effect of helium implantation on mechanical properties of EUROFER97 evaluated by nanoindentation. *J. Nucl. Mater.* **2014**, *448*, 301–309. [CrossRef]
16. Lucas, G.E. The evolution of mechanical property change in irradiated austenitic stainless steels. *J. Nucl. Mater.* **1993**, *206*, 287–305. [CrossRef]
17. Gan, J.; Was, G.S. Microstructure evolution in austenitic Fe–Cr–Ni alloys irradiated with rotons: Comparison with neutron-irradiated microstructures. *J. Nucl. Mater.* **2001**, *297*, 161–175. [CrossRef]
18. Xiao, X.; Chen, Q.; Yang, H.; Duan, H.; Qu, J. A mechanistic model for depth-dependent hardness of ion irradiated metals. *J. Nucl. Mater.* **2017**, *485*, 80–89. [CrossRef]

19. Chen, S.; Wang, Y.; Hashimoto, N.; Ohnuki, S. Post-irradiation annealing behavior of helium in irradiated Fe and ferritic-martensitic steels. *Nuclear Mater. Energy* **2018**, *15*, 203–207. [CrossRef]
20. Federici, G.; Bachmann, C.; Biel, W.; Boccaccini, L.; Cismondi, F.; Ciattaglia, S.; Coleman, M.; Day, C.; Diegele, E.; Franke, T.; et al. Overview of the design approach and prioritization of R&D activities towards an EU DEMO. *Fusion Eng. Des.* **2016**, *109–111 Pt B*, 1464–1474. [CrossRef]
21. Stork, D.; Agostini, P.; Boutard, J.-L.; Buckthorpe, D.; Diegele, E.; Dudarev, S.L.; English, C.; Federici, G.; Gilbert, M.R.; Gonzalez, S.; et al. Materials R&D for a timely DEMO: Key findings and recommendations of the EU Roadmap Materials Assessment Group. *Fusion Eng. Des.* **2014**. [CrossRef]
22. Gruber, E.E. Calculated Size Distributions for Gas Bubble Migration and Coalescence in Solids. *J. Appl. Phys.* **1967**, *38*, 243–250. [CrossRef]
23. Russell, K.C. Nucleation of voids in irradiated metals. *Acta Metall.* **1971**, *19*, 753–758. [CrossRef]
24. Greenwood, G.W.; Boltax, A. The role of fission gas re-solution during post-irradiation heat treatment. *J. Nucl. Mater.* **1962**, *5*, 234–240. [CrossRef]
25. Markworth, A. On the coarsening of gas-filled pores in solids. *Metall. Trans.* **1973**, *4*, 2651–2656. [CrossRef]
26. Ostwald, W. Zur Energetik. *Ann. Phys.* **1896**, *294*, 154–167. [CrossRef]
27. Roldán, M.; Fernández, P.; Rams, J.; Jiménez-Rey, D.; Materna-Morris, E.; Klimenkov, M. Comparative study of helium effects on EU-ODS EUROFER and EUROFER97 by nanoindentation and TEM. *J. Nucl. Mater.* **2015**, *460*, 226–234. [CrossRef]
28. Reduced Activation Ferritic/Martensitic Steel Eurofer'97 as Possible Structural Material for Fusion Device. Metallurgical Characterization on As-Received Condition and after Simulated Service Conditions. Available online: https://inis.iaea.org/collection/NCLCollectionStore/_Public/36/026/36026402.pdf (accessed on 29 November 2018).
29. Sandim, H.R.Z.; Renzetti, R.A.; Padilha, A.F.; Raabe, D.; Klimenkov, M.; Lindau, R.; Möslang, A. Annealing behavior of ferritic–martensitic 9%Cr–ODS–Eurofer steel. *Mater. Sci. Eng. A* **2010**, *527*, 3602–3608. [CrossRef]
30. Robinson, M.T. Slowing-down time of energetic atoms in solids. *Phys. Rev. B* **1989**, *40*, 10717–10726. [CrossRef]
31. Robinson, M.T. Computer simulation studies of high-energy collision cascades. *Nucl. Instrum. Methods Phys. Res. Sect. B Beam Interact. Mater. At.* **1992**, *67*, 396–400. [CrossRef]
32. Oliver, W.C.; Pharr, G.M. An improved technique for determining hardness and elastic modulus using load and displacement sensing indentation experiments. *J. Mater. Res.* **1992**, *7*, 29. [CrossRef]
33. Oliver, W.C.; Pharr, G.M. Measurement of hardness and elastic modulus by instrumented indentation: Advances in understanding and refinements to methodology. *J. Mater. Res.* **2004**, *19*, 3–20. [CrossRef]
34. Fischer-Cripps, A.C.; SpringerLink. *Nanoindentation*; Springer: New York, NY, USA, 2011.
35. Krier, J.; Breuils, J.; Jacomine, L.; Pelletier, H. Introduction of the real tip defect of Berkovich indenter to reproduce with FEM nanoindentation test at shallow penetration depth. *J. Mater. Res.* **2012**, *27*, 28–38. [CrossRef]
36. Tan, L.; Byun, T.S.; Katoh, Y.; Snead, L.L. Stability of MX-type strengthening nanoprecipitates in ferritic steels under thermal aging, stress and ion irradiation. *Acta Mater.* **2014**, *71*, 11–19. [CrossRef]
37. Jiang, C.; Swaminathan, N.; Deng, J.; Morgan, D.; Szlufarska, I. Effect of Grain Boundary Stresses on Sink Strength. *Mater. Res. Lett.* **2013**, *2*, 100–106. [CrossRef]
38. Allen, S.M. Foil thickness measurements from convergent-beam diffraction patterns. *Philos. Mag. A* **1981**, *43*, 325–335. [CrossRef]
39. Allen, S.M.; Hall, E.L. Foil thickness measurements from convergent-beam diffraction patterns An experimental assessment of errors. *Philos. Mag. A* **1982**, *46*, 243–253. [CrossRef]
40. Glazer, J.; Ramesh, R.; Hilton, M.R.; Sarikaya, M. Comparison of convergent-beam electron diffraction methods for determination of foil thickness. *Philos. Mag. A* **1985**, *52*, L59–L63. [CrossRef]
41. Kirk, M.; Yi, X.; Jenkins, M. Characterization of irradiation defect structures and densities by transmission electron microscopy. *J. Mater. Res.* **2015**, *30*, 1195–1201. [CrossRef]
42. Yao, B.; Edwards, D.J.; Kurtz, R.J.; Odette, G.R.; Yamamoto, T. Multislice simulation of transmission electron microscopy imaging of helium bubbles in Fe. *J. Electron Microsc.* **2012**, *61*, 393–400. [CrossRef] [PubMed]
43. Jenkins, M.L.; Kirk, M.A. *Characterization of radiation damage by TEM*; Institute of Physics: London, UK, 2001.
44. Xia, L.D.; Liu, W.B.; Liu, H.P.; Zhang, J.H.; Chen, H.; Yang, Z.G.; Zhang, C. Radiation damage in helium ion–irradiated reduced activation ferritic/martensitic steel. *Nucl. Eng. Technol.* **2018**, *50*, 132–139. [CrossRef]

45. Gao, J.; Liu, Z.-J.; Wan, F.-R. Limited Effect of Twin Boundaries on Radiation Damage. *Acta Metall. Sin. (Engl. Lett.)* **2016**, *29*, 72–78. [CrossRef]

46. Singh, B.N. Effect of grain size on void formation during high-energy electron irradiation of austenitic stainless steel. *Philos. Mag. A J. Theor. Exp. Appl. Phys.* **1974**, *29*, 25–42. [CrossRef]

47. Was, G.S. *Fundamentals of Radiation Materials Science (Metals and Alloys)*; Springer: Berlin, Germany, 2007.

48. Nagasaka, T.; Shibayama, T.; Kayano, H.; Hasegawa, A.; Satou, M.; Abe, K. Effect of purity on helium bubble formation in 9Cr martensitic steel during post-implantation annealing at 1105 K. *J. Nucl. Mater.* **1998**, *258–263 Pt 2*, 1193–1198. [CrossRef]

49. Pouchon, M.A.; Chen, J.; Döbeli, M.; Hoffelner, W. Oxide dispersion strengthened steel irradiation with helium ions. *J. Nucl. Mater.* **2006**, *352*, 57–61. [CrossRef]

50. Gai, X.; Lazauskas, T.; Smith, R.; Kenny, S.D. Helium bubbles in bcc Fe and their interactions with irradiation. *J. Nucl. Mater.* **2015**, *462*, 382–390. [CrossRef]

51. Luo, F.; Guo, L.; Chen, J.; Li, T.; Zheng, Z.; Yao, Z.; Suo, J. Damage behavior in helium-irradiated reduced-activation martensitic steels at elevated temperatures. *J. Nucl. Mater.* **2014**, *455*, 339–342. [CrossRef]

52. Trinkaus, H. The effect of cascade induced gas resolution on bubble formation in metals. *J. Nucl. Mater.* **2003**, *318*, 234–240. [CrossRef]

53. Foreman, A.J.E.; Singh, B.N. The role of collision cascades and helium atoms in cavity nucleation. *Radiat. Eff. Defects Solids* **1990**, *113*, 175–194. [CrossRef]

54. Dethloff, C. *Modeling of Helium Bubble Nucleation and Growth in Neutron Irradiated RAFM Steels*; KIT Scientific Publishing: Karlsruhe, Germany, 2012.

55. Trinkaus, H. The effect of internal pressure on the coarsening of inert gas bubbles in metals. *Scr. Metall.* **1989**, *23*, 1773–1778. [CrossRef]

56. Ono, K.; Arakawa, K.; Hojou, K. Formation and migration of helium bubbles in Fe and Fe–9Cr ferritic alloy. *J. Nucl. Mater.* **2002**, *307–311 Pt 2*, 1507–1512. [CrossRef]

57. Borodin, V.A.; Vladimirov, P.V. Kinetic properties of small He–vacancy clusters in iron. *J. Nucl. Mater.* **2009**, *386–388*, 106–108. [CrossRef]

58. Nagasaka, T.; Shibayama, T.; Kayano, H.; Hasegawa, A.; Abe, K. Microstructures of high-purity ferritic steels after helium implantation. *Sci. Rep. Rerearch Inst. Tohoku Univ. Ser. A-Phys.* **1997**, *45*, 121–126.

59. Nagasaka, T.; Shibayama, T.; Kayano, H.; Hasegawa, A.; Abe, K. Effects of several impurity additions on microstructural evolution in high-purity Fe-9Cr ferritic alloys after helium implantation. *Phys. Status Solidi A Appl. Res.* **1998**, *167*, 335–346. [CrossRef]

60. Ishiyama, Y.; Kodama, M.; Yokota, N.; Asano, K.; Kato, T.; Fukuya, K. Post-irradiation annealing effects on microstructure and helium bubbles in neutron irradiated type 304 stainless steel. *J. Nucl. Mater.* **1996**, *239*, 90–94. [CrossRef]

61. Stoller, R.E.; Odette, G.R. The effects of helium implantation on microstructural evolution in an austenitic alloy. *J. Nucl. Mater.* **1988**, *154*, 286–304. [CrossRef]

62. Maziasz, P.J. Effects of Helium Content on Microstructural Development in Type 316 Stainless Steel Under Neutron Irradiation. Ph. D. Thesis, University of Tennessee, Knoxville, TN, USA, 1984.

63. Golubov, S.I.; Stoller, R.E.; Zinkle, S.J.; Ovcharenko, A.M. Kinetics of coarsening of helium bubbles during implantation and post-implantation annealing. *J. Nucl. Mater.* **2007**, *361*, 149–159. [CrossRef]

64. Fu, C.-C.; Torre, J.D.; Willaime, F.; Bocquet, J.-L.; Barbu, A. Multiscale modelling of defect kinetics in irradiated iron. *Nat. Mater.* **2005**, *4*, 68–74. [CrossRef]

65. Brodrick, J.; Hepburn, D.J.; Ackland, G.J. Mechanism for radiation damage resistance in yttrium oxide dispersion strengthened steels. *J. Nucl. Mater.* **2014**, *445*, 291–297. [CrossRef]

66. Lazauskas, T.; Kenny, S.D.; Smith, R.; Nagra, G.; Dholakia, M.; Valsakumar, M.C. Simulating radiation damage in a bcc Fe system with embedded yttria nanoparticles. *J. Nucl. Mater.* **2013**, *437*, 317–325. [CrossRef]

micromachines

MDPI

Article

Static and Fatigue Tests on Cementitious Cantilever Beams Using Nanoindenter

Yidong Gan, Hongzhi Zhang *, Branko Šavija, Erik Schlangen and Klaas van Breugel

Microlab, Faculty of Civil Engineering and Geosciences, Delft University of Technology, 2628 CN Delft, The Netherlands; y.gan@tudelft.nl (Y.G.); B.Savija@tudelft.nl (B.Š.); Erik.Schlangen@tudelft.nl (E.S.); K.vanBreugel@tudelft.nl (K.v.B.)

* Correspondence: H.Zhang-5@tudelft.nl; Tel.: +31-685-515-349

Received: 2 November 2018; Accepted: 22 November 2018; Published: 28 November 2018

check for updates

Abstract: Cement paste is the main binding component in concrete and thus its fundamental properties are of great significance for understanding the fracture behaviour as well as the ageing process of concrete. One major aim of this paper is to characterize the micromechanical properties of cement paste with the aid of a nanoindenter. Besides, this paper also presents a preliminary study on the fatigue behaviour of cement paste at the micrometer level. Miniaturized cantilever beams made of cement paste with different water/cement ratios were statically and cyclically loaded. The micromechanical properties of cement paste were determined based on the measured load-displacement curves. The evolution of fatigue damage was evaluated in terms of the residual displacement, strength, and elastic modulus. The results show that the developed test procedure in this work is able to produce reliable micromechanical properties of cement paste. In addition, little damage was observed in the cantilever beams under the applied stress level of 50% to 70% for 1000 loading cycles. This work may shed some light on studying the fatigue behaviour of concrete in a multiscale manner.

Keywords: cement paste; miniaturized cantilever beam; micromechanics; fatigue; nanoindenter

1. Introduction

Concrete is the most widely used construction material in the world [1]. As a heterogeneous material, concrete is mainly made of cement, aggregate, and water. It is well acknowledged that the fracture of concrete is a complex phenomenon due to its innate multiphase and multiscale heterogeneity [2]. As a result, the multiscale modelling method is generally adopted to investigate the fracture behaviour of concrete at different length scales [2,3]. By explicitly considering the material structures and material properties of individual constituents at the finer scale, the macroscopic mechanical properties of concrete can be predicted. Therefore, as one of the primary inputs for the multiscale approach, the local mechanical properties of constitutes are needed.

In particular, the cement paste serving as a major component is of great significance with respect to the fracture process of concrete. At the micro level, it is generally recognized that the cement paste includes several different phases: Pores, low density calcium-silicate-hydrate (C-S-H), high density C-S-H, and anhydrous clinker minerals [4,5]. During the last decade, many efforts have been devoted to characterizing the micromechanical properties of cement paste [5–9]. Among them, nanoindentation is a commonly applied method for determination of the local elastic modulus and hardness [5,6]. However, one major concern of this technique is that the local strength of materials is indirectly obtained from the measured hardness, while the correlation between the hardness and strength of cement paste has not been experimentally examined so far [10]. Furthermore, the local mechanical properties of cement paste obtained from nanoindentation tests are largely affected by the spatial heterogeneity of the material [10]. Therefore, the necessity of a more robust test method to determine the micromechanical properties of cement paste is addressed. Recently, Schlangen et al. [11] and Zhang et al. [12] developed a new test procedure to prepare small cement paste cubes (100 μm) with the application of a precision micro-dicing instrument and then to investigate their structural response using a nanoindenter. This promising work successfully provides the experimental evidence for calibration and validation of multiscale modelling techniques. Moreover, to better understand the fracture behaviour of cement paste at the microscale, multiple types of tests should be conducted. Therefore, this concept of small scale testing on cement paste has also been extended to perform splitting tests [13] and three-point bending tests [14]. Similarly, Němeček et al. [9] also conducted small scale bending tests on cantilever micro-beams (20 μm) fabricated by focused ion beam technology and the tensile strength of an individual component in the cement paste was directly measured. However, a major disadvantage of this method is the time-consuming fabrication process of the specimen, which results in a small number of specimens that can be tested. It should be noted that at the microscale, a large scatter of the mechanical properties of cement paste can be expected [15,16]. It means that a sufficient number of tests should be performed to guarantee the statistical reliability [13]. In this paper, the beams with a square cross-section of 380 μm × 380 μm were fabricated by a micro dicing saw. The mechanical properties of cement paste at the microscale were measured by conducting bending tests on miniaturized cantilever beams with the aid of a nanoindenter. The main benefits of this approach are that the mechanical properties can be reliably determined in an easy and straightforward manner. Furthermore, a sufficient number of samples can be tested in a relatively short period of time.

Different from the fracture behaviour under monotonic loading, fatigue fracture of concrete is a process of progressive, permanent internal structural changes, which inevitably result in the changes of the performance of concrete with the elapse of time. This ageing phenomenon of concrete is mainly attributed to the growth of internal microcracks, which eventually coalesce into macrocracks and lead to complete fracture after a sufficient number of cycles [17,18]. Meanwhile, the time-dependent growth of microcracks will accelerate the formation of different gradients in concrete [19,20], and hence, further increase the proneness of concrete to ageing [21]. Although numerous studies have been conducted on the fatigue behaviour of concrete, most of them deal with the global approaches by means of S-N curves or fracture mechanics [22–25]. The common shortcoming of them is that the inherent heterogeneity of concrete is neglected, and thus the realistic behaviour of concrete under fatigue loading, e.g., the propagation of microcracks, is difficult to be predicted. In order to gain a better understanding of the fatigue fracture of concrete, the multiscale approach is expected to be a proper choice [26–28]. According to the study of [27], a mesoscopic model based on the real microstructure was established to investigate the evolution of fatigue damage in concrete, in which the different fatigue damage functions of individual components were assumed. However, the independent fatigue behaviour of different components has never been tested experimentally. To fill this gap, a preliminary study on the fatigue behaviour of cement paste at the micrometre length scale was also carried out in this paper.

2. Materials and Methods

2.1. Materials and Sample Preparation

In this study, standard grade CEM I 42.5 N Portland cement and deionized water were used to prepare the cement paste. The cement was mixed with water to yield three different water/cement ratios (0.3, 0.4, and 0.5). After mixing, the mixture was cast in plastic cylindrical moulds with a 24 mm diameter and 39 mm height. In order to reduce the amount of entrapped air, the fresh paste was carefully compacted and stirred layer by layer on a vibrating table. Afterwards, the sample was covered with a plastic foil and rotated at a speed of 2.5 rpm at room temperature (20 °C) for 24h, aiming to mitigate the influence of bleeding. All specimens were cured in a sealed condition at room temperature for 28 days. After that, specimens were demoulded and then cut into slices with a thickness of 3 mm. The solvent exchange method was adopted to arrest the hydration reactions of samples by immersing them in isopropanol [29]. A detailed procedure of arresting hydration can be found in [14].

For the fabrication process of miniaturized cantilever beams, several steps were followed: Firstly, the two ends of slices were further ground using a Struers Labopol-5 thin sectioning machine. To obtain smooth and parallel surfaces, grinding discs with two different grit sizes of 135 μm and 35 μm were used in sequence. The final thickness of the sample was approximately 2.15 mm; the next step is to generate the miniaturized cantilever beams from the thin samples. This is achieved by utilizing a precision micro-dicing machine (MicroAce Series 3 Dicing Saw, Loadpoint Limited, Swindon, UK), which is mainly applied to cut semiconductor wafers. By applying two perpendicular cutting directions and same cutting space, a row of cantilever beams, including at least 20–30 beams, with a square cross section of 380 μm × 380 μm were obtained. The cantilevered length was approximately 1.65 mm and the thickness of the baseplate was less than 500 μm. The cutting process is schematically shown in Figure 1. Afterwards, the cross-section of the beams was examined by using an environmental scanning electron microscope (ESEM). The fabrication process yields an overall accuracy for the cross-sectional dimensions of ±5 μm (Figure 2b).

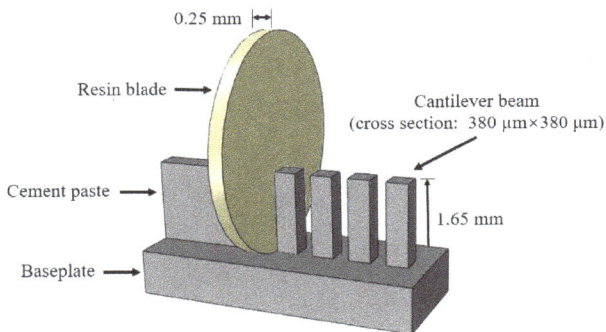

Figure 1. Schematic diagram of sample preparation.

Figure 2. Environmental scanning electron microscope (ESEM) image of: (**a**) A row of cantilever beams; (**b**) cross-section of a cantilever beam.

2.2. Static Bending Tests

For each w/c ratio, a total of 30 cantilever beams were tested under static bending. An Agilent G200 nanoindenter operating with a continuous stiffness measurement (CSM) function was selected to conduct the static bending tests. The CSM technique allows for the continuous measurement of mechanical properties of materials, thus eliminating the need for unloading cycles [30]. In this study, an oscillating force with a harmonic displacement of 2 nm at a frequency of 45 Hz was adopted for the CSM method. The set-up of the static test is illustrated in Figure 3. A nut was first fixed on a metal block, which acts as the supporting base. Then a row of cantilever beams was horizontally attached on the flat wall of the fixed nut using cyanoacrylate adhesive (Loctite Superglue 3). A flat end cylindrical diamond indenter with a diameter of 315 μm was used instead of the mostly used Berkovich indenter (a sharp pyramid tip) for the standard nanoindentation test [11,31,32]. In this way, the possible penetration of the indenter into the specimen was eliminated. The built-in imaging technique of the nanoindenter was used to accurately locate the load point at the central axis of the beams. The load point was always kept around 150 μm from the free end of the beam. The loading procedure was displacement controlled with a 50 nm·s^{-1} loading rate. The cantilever beams were loaded until failure and the load-displacement curves were collected by the nanoindenter. After failure, the distance between the load point and the fracture point on the cantilever beam was measured for the calculation of the stress.

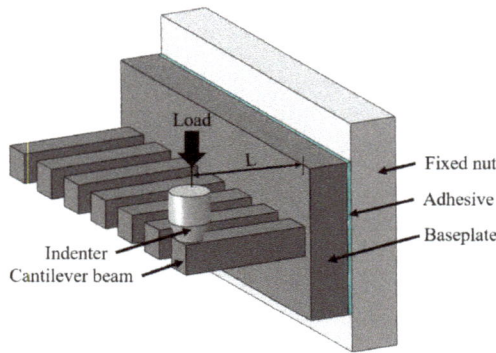

Figure 3. Schematic diagram of test set-up.

2.3. Fatigue Bending Tests

Fatigue tests were performed using the same experimental apparatus (Agilent G200 nanoindenter) as static tests. In total, nine cantilever beams with a w/c ratio of 0.5 were tested under the same upper and lower load. The tests were load-controlled with a constant upper load of 100 mN. The ratio, R, between the minimal and maximum load was 0.05. Figure 4 illustrates a schematic diagram of the load-cycle curve used in the fatigue tests. The tests were carried out using the constant amplitude triangular load with a frequency of 0.33 Hz. Note that due to the material heterogeneity, the realistic stress level (tensile stress divided by the static strength) for each specimen is varied even under the same maximum load. For the sake of simplicity, the average value of 63% is adopted to describe the applied stress level. Due to the limitation of test duration in the nanoindenter, four identical load blocks with 250 loading cycles (in total 1000 cycles) were performed on each beam and all of them were conducted in succession without any rest period. Since the fatigue load must be unloaded to zero after completing each load block, the residual displacement was recorded to connect with the succeeding block. Besides, before the end of each load block, the indenter was held for around 50 s at a lower load for the purpose of thermal drift correction [33]. After reaching the target number of cycles, the beams were statically loaded until failure. The load-displacement curves for fatigue tests as well as static tests were recorded by the nanoindenter.

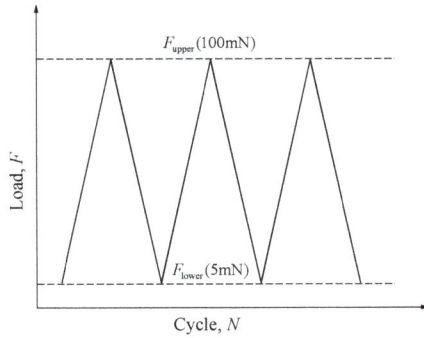

Figure 4. Schematic diagram of the load-cycle curve.

3. Results and Discussions

3.1. Static Bending Tests

The load-displacement curves for different w/c ratios are presented in Figure 5. A high degree of repeatability is found for measurements on 30 cantilever beams with the same w/c ratio. It can be seen that the load-displacement curves show two distinct stages. In the first stage, the displacement increases linearly with the increase of load until the maximum load is reached. It is then followed by a rapid burst of displacement in the second stage, where a nearly brittle fracture occurred. Since the displacement control of the nanoindenter is not fast enough to capture the post-peak behaviour of the specimen, it results in an overshoot of the indention tip, thus presenting a horizontal line in the load-displacement curve [34].

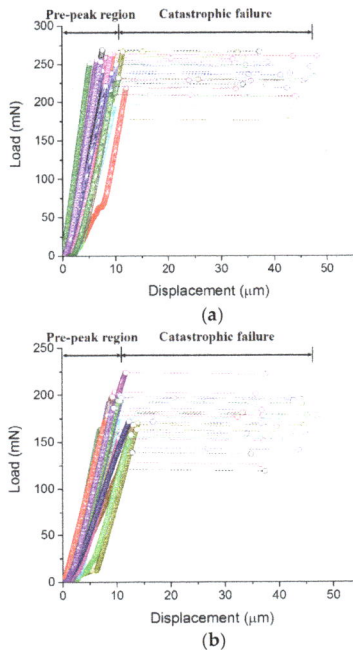

(a)

(b)

Figure 5. *Cont.*

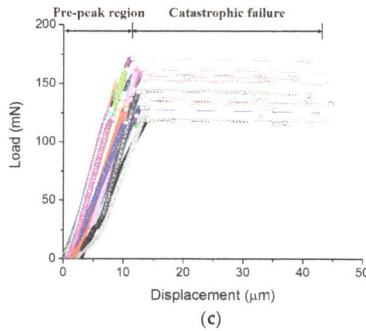

(c)

Figure 5. Load-displacement curves of beams with a w/c ratio of (**a**) 0.3; (**b**) 0.4; and (**c**) 0.5.

Figure 6 shows the ESEM images of cantilever beams before and after the bending test. It appears that the tested beam has a rough fracture surface and may be resulted from the randomly distributed defects (pores). It should be noted that the cracks usually initiate at the weakest point near the fixed end of the cantilever beams and eventually lead to the complete fracture.

Figure 6. ESEM images of beams (**a**) before fracture; (**b**) after fracture

The load-displacement curve data is used to estimate the mechanical properties of the cement paste, i.e., the strength and elastic modulus. For static beam experiments, the strength is commonly defined as the tensile stress in the failure plane extreme fiber under maximum load, F_{max} [35]. Assuming the beam is loaded without torsion, the strength of beams, f_t, is given by:

$$f_t = \frac{F_{max}d}{I_y}\frac{h}{2} \tag{1}$$

where d is the distance between the load point and the fracture point, h is the side length of the square cross-section, and $I_y = h^4/12$ is the moment of inertia. For the calculation of the elastic modulus, E, the linear region of the load-displacement curve is used. Note that the effects of shear deformation can be negligible only when the length to depth of the beam is large enough (normally $\geq$5.0) [9,36,37], which is, however, not the case in this study (around 4.3). Therefore, according to the Timoshenko beam theory [35], the elastic modulus in consideration of the shear effect can be computed as:

$$E = \frac{L^3}{3I_y}k + \frac{2(1+\nu)L}{\kappa L^2}k \tag{2}$$

where L is the effective length of the cantilever beam; k is the beam stiffness, which is defined as the slope in the linear region of the load-displacement curve; and v and κ are the Poisson's ratio of the cement paste and shear coefficient for rectangular cross-section, respectively. Considering the values reported in the literature [3,7,38–40], $v = 0.25$ and $\kappa = 5/6$ are used in this study. In addition, it should be borne in mind that the measured load-displacement curves also include the influence of the baseplate and adhesive, which may potentially underestimate the stiffness of the beam [36]. To evaluate the effects of the baseplate and adhesive on the overall stiffness, a finite element model using commercial software, ABAQUS, was established. The model consists of approximately 9000, three dimensional, eight-node brick elements (C3D8R), with a geometry as shown in Figure 3. In this model, the thickness of the adhesive was assumed to be 50 μm, and the elastic moduli of the cement paste and adhesive were set to 15 GPa and 3 GPa, respectively. The thickness of the baseplate used in the model varied from 100 μm to 500 μm. Meanwhile, a single cantilever beam with a fixed end was also simulated for comparison. An example of the numerical model, including the mesh and boundary conditions, is illustrated in Figure 7. The simulation results reveal that the additional deflection caused by the adhesive is not more than 0.5–0.6% of the total beam deflection, which is thought to be negligible. However, the extra deflection caused by the baseplate accounts for around 5–20% of the total beam deflection, largely depending on the thickness of baseplate. For the sake of simplicity, in this study, the extra deflections caused by the adhesive and baseplate are not taken into account in the estimate of the elastic modulus. Nevertheless, continuous improvement of the test procedure is required to further minimize the discrepancies in measured material properties, which is not attributed to the materials themselves.

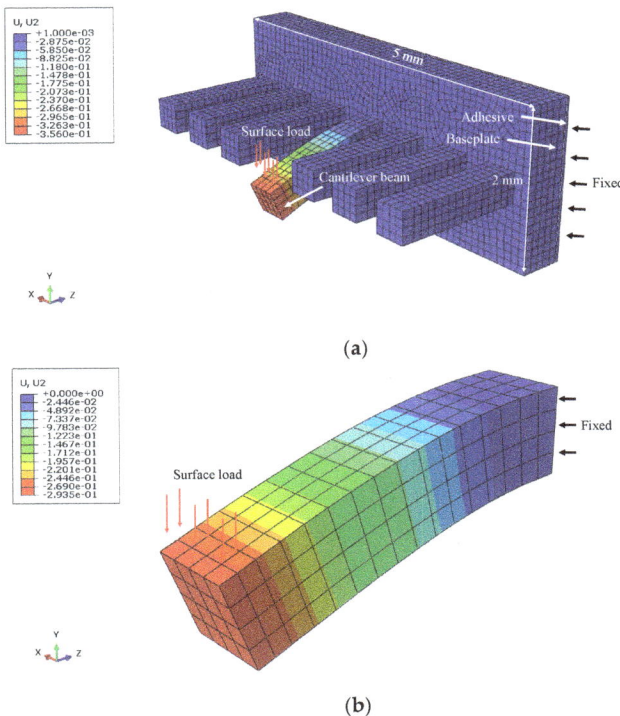

(a)

(b)

Figure 7. The elastic analysis of (**a**) the cantilever beam with the baseplate and adhesive; (**b**) a single cantilever beam.

The calculated mean value and standard deviation of the strength and elastic modulus are summarized in Table 1, together with the data reported in the literature [14] (in brackets). As expected, both the strength and elastic modulus decrease with an increasing w/c ratio. It can also be seen that the calculated elastic modulus agrees well with the literature results mentioned by [14], while the calculated strength is almost 1.5–2.0 times higher than the results in the literature. It is believed that the longer beam (i.e., 12 mm) used in [14] is the major reason for the discrepancy in the calculated strength despite the possible influence of different boundary conditions. The observed decrease in strength for larger samples is a common feature for quasi-brittle material and can be explained by the size effect [41–43].

Table 1. Calculated mechanical properties, compared with previous work [14].

w/c Ratio	Strength (MPa)	Elastic Modulus (GPa)
0.3	31.34 ± 3.70 (20.28) [1]	18.85 ± 3.23 (16.68) [1]
0.4	25.27 ± 3.23 (15.31) [1]	13.97 ± 1.98 (12.79) [1]
0.5	22.37 ± 2.31 (11.71) [1]	10.80 ± 2.44 (9.09) [1]

[1] The value in brackets are from the reference [14].

3.2. Fatigue Bending Tests

3.2.1. Load-Displacement Curve

The typical measured load-displacement curve of the fatigue test is shown in Figure 8. The figure shows that the load-displacement relation for each cycle is approximately linear, and also the residual displacement are accumulated with the increase of loading cycles. The variation of stiffness during loading characterized by the slope of the loading curve for all fatigue tests are plotted in Figure 9, together with the line of best fit. It is found that the loading stiffness changes very little with the number of cycles, which is associated with a slow accumulation of damage during the fatigue loading. It should be noted that the fatigue response of material is very sensitive to the flaws and defects [44–46]. Consequently, when compared with the mortar or concrete at the larger scale, the cement paste at the micro scale exhibits higher fatigue resistance due to less initial defects, such as the porous interfacial transition zone (ITZ) [47].

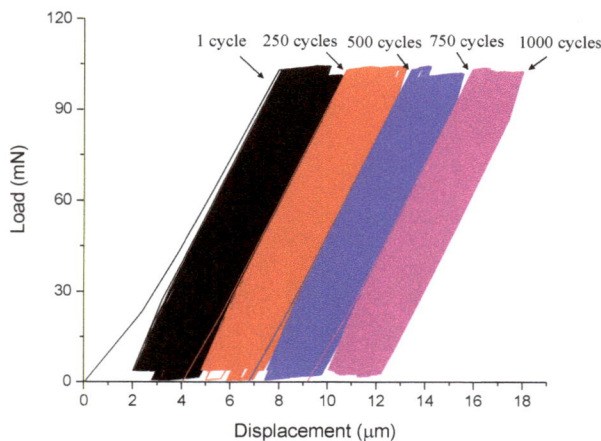

Figure 8. A typical load-displacement curve for the fatigue test.

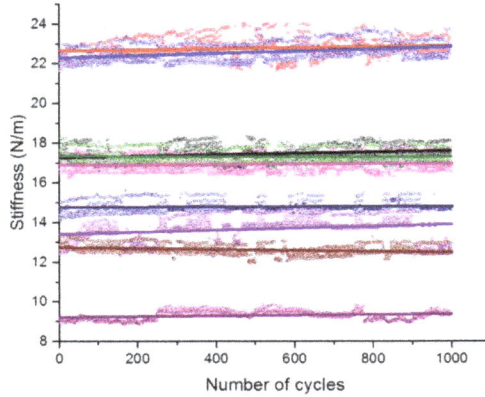

Figure 9. The loading stiffness versus the number of cycles for all fatigue tests (different colors represent different specimens).

Although no evident decline of loading stiffness can be identified in all fatigue tests, the relation between the displacement and number of cycles exhibits a typical fatigue damage evolution curve [18,48–50], see Figure 10a. The increasing displacement under each loading block can be divided into a transient primary stage and a steady secondary stage. It can also be seen from Figure 10b that the growth rate of displacement decreases rapidly in the primary stage and reaches a constant value, with relatively small fluctuations in the secondary stage. These two stages are mainly related to the formation and progressive growth of internal microcracks in cementitious materials, respectively [24,51].

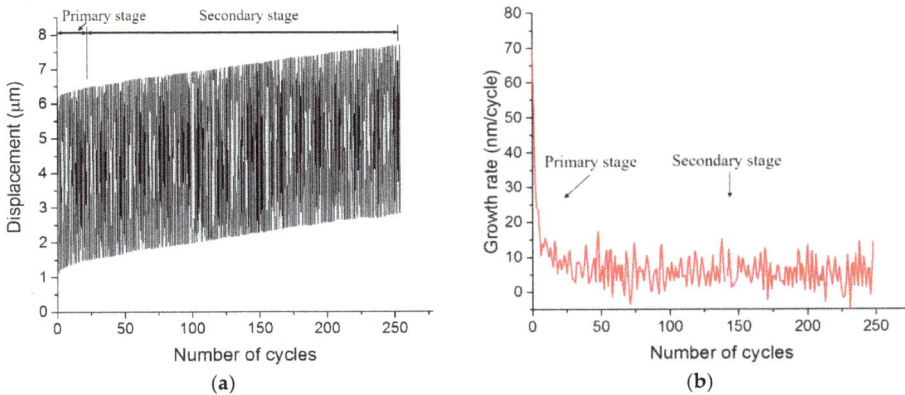

Figure 10. A typical curve of (**a**) displacement versus the number of cycles; (**b**) growth rate of displacement with cycles.

3.2.2. Growth Rate of Residual Displacement

As an important indicator of the fatigue damage, the residual displacements at a lower load are extracted from the load-displacement curve. A typical curve for the development of residual displacement under four identical loading blocks is plotted in Figure 11. Despite the discontinuity between each load block, the development of residual displacement shows an overall increasing trend with nearly the same slope. This indicates a stable propagation of microcracks in cantilever beams even with the interruption of cyclic loading. Figure 12 presents the constant growth rate of the residual displacement for each load block as well as the average value of all tests (6.47 ± 1.01 nm/cycle). Note that the data with large deviations are excluded in the calculation of the average value. Figure 12 indicates that the growth rate of each load block shows a high degree of consistency. Nevertheless, the value between each specimen is relatively scattered, mainly due to the natural heterogeneity of cementitious materials.

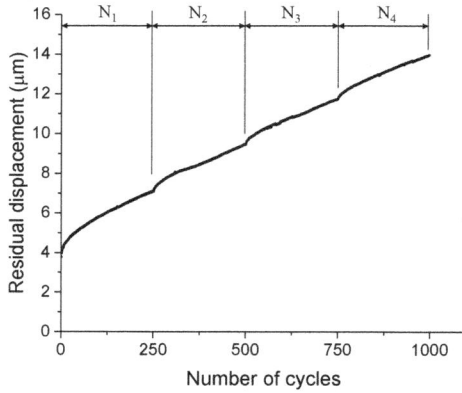

Figure 11. A typical curve for the development of residual displacement (N_1, N_2, N_3, and N_4 denote the four load blocks).

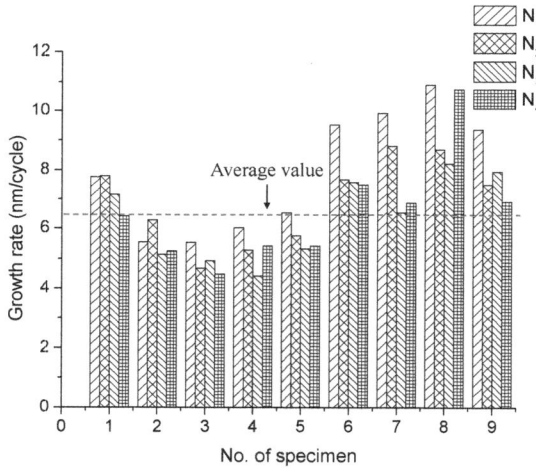

Figure 12. The growth rate of residual displacement for all specimens (N_1, N_2, N_3, and N_4 denote the four load blocks).

3.2.3. Residual Mechanical Properties

To quantitatively assess the degree of fatigue damage, all damaged cantilever beams were loaded statically to failure. The percentage reduction of the strength and elastic modulus for each beam were plotted against the stress level in Figure 13. The obtained average percentage reduction of the strength and elastic modulus are 15.29% and 15.78%, respectively.

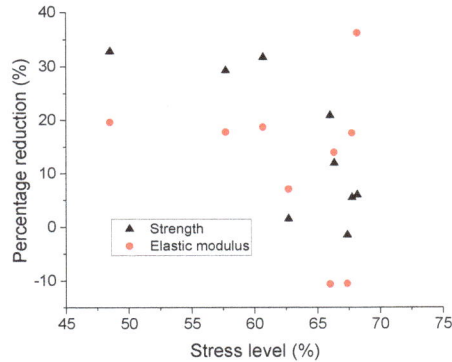

Figure 13. The relation between the percentage reduction of mechanical properties and the stress level.

As expected, a large scatter can be observed in this diagram. For instance, the percentage reduction of strength ranges from 1% to 20% even under the same stress level of 67%, while the percentage reduction of the elastic modulus lies in a wider range of −10% to 36%. A downtrend inferred from Figure 13 indicates that the higher stress level leads to the lower percentage reduction of mechanical properties, which seems to be unrealistic. On the other hand, since the number of tests is relatively small, it is not enough to make a rational judgement of the results. Nevertheless, the observed scatter in Figure 13 implies that the applied stress level may not be a proper parameter to correlate the residual mechanical properties at this length scale. The cause of the failed interpretation of the results is mainly attributed to the assumption of a constant strength. However, cement paste at the micro scale contains different size of pores and initial defects [5,7,52]. Most of them are formed during the hydration process [53] or induced by the external factors, such as casting and vibration. For instance, a large air void can be detected in the cross-section of the cantilever beam, as shown in Figure 14. Therefore, the randomly distributed pores may result in the difference of original strength. This is consistent with previous numerical work [52], in which the predicted mechanical properties of cement paste at the microscale were largely influenced by the real microstructures.

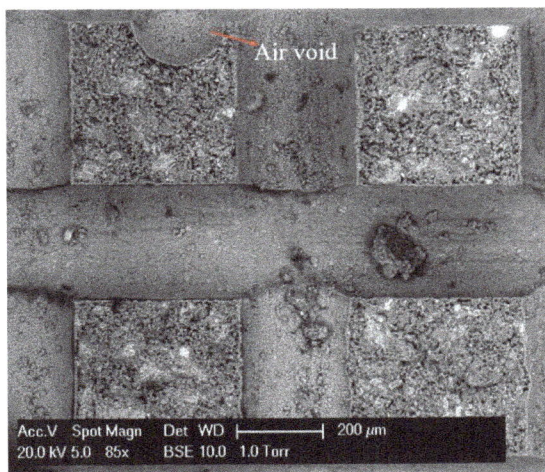

Figure 14. The ESEM image of a large air void detected in the fractured cross-section of cantilever beams.

To consider the individual variation of strength of each specimen, the ratio of applied stress to residual strength (denoted as R_0) is adopted to evaluate the fatigue damage. The relation between the residual mechanical properties and the ratio, R_0, are shown in Figure 15. A clear trend can be found that the percentage reduction of strength and elastic modulus increase with the increase of the ratio, R_0. It indicates that the employed parameter is likely to give a proper analysis of the data, but this only applies to the beams undergoing low fatigue damage. It should also be kept in mind that to acquire a reliable prediction of fatigue behaviour of the cement paste, a large number of tests is needed.

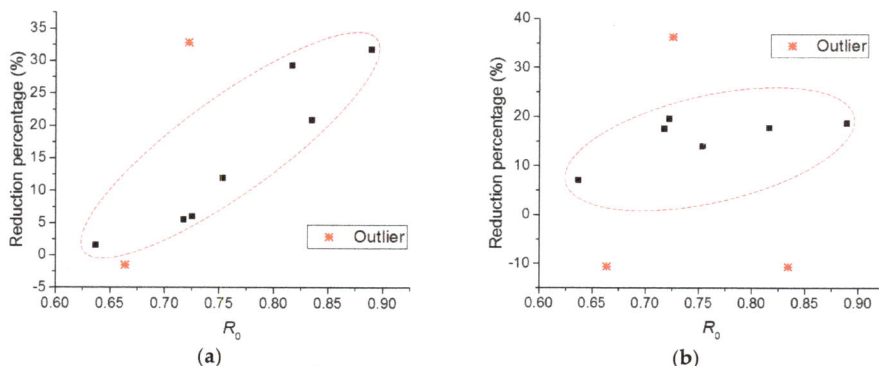

Figure 15. The relation between the percentage reductions of (**a**) strength; (**b**) elastic modulus against the ratio, R_0.

4. Conclusions

In this study, the nanoindenter technique was used to investigate the micromechanical properties of cement paste. This work also represents the first attempt in studying the fatigue behaviour of cement paste at the micrometer length scale. From the results of the current study, we can conclude that the employed test procedure is able to obtain reliable mechanical properties of cement paste at the micro scale. In addition, the cement paste at the microscale exhibits higher fatigue resistance due to less initial damage. The fatigue damage evolution can be quantitatively assessed in terms of the residual displacement and the residual mechanical properties. In consideration of the heterogeneous microstructure of cement paste, the ratio of stress to residual strength is suggested to properly analyze the evolution of fatigue damage. This study provides new insights into the fatigue behaviour of cement paste, which may shed some light for future studies on the multiscale modelling of the ageing process of concrete subjected to fatigue loading. Systematic and additional studies are required to test the effect of the w/c ratio, loading history, and rest period on the fatigue behaviour of cement paste at the microscale. More importantly, the fatigue behaviour of ITZ at this scale needs to be further investigated.

Author Contributions: Conceptualization, Y.G., H.Z., B.Š., E.S. and K.v.B.; Methodology, Y.G., H.Z., B.Š., E.S. and K.v.B.; Software, Y.G.; Data Curation, Y.G. and H.Z.; Writing—Original Draft Preparation, Y.G.; Writing—Review & Editing, Y.G., H.Z., B.Š., E.S. and K.v.B.; Supervision, B.Š., E.S. and K.v.B.

Funding: This research was funded by China Scholarship Council (CSC) under grant No. 201706130140 and CSC No. 201506120067.

Acknowledgments: The authors would like to acknowledge the help of Arjan Thijssen with ESEM experiments.

Conflicts of Interest: The authors declare no conflict of interest.

References

1. Mehta, P.K.; Monteiro, P.J.M. *Concrete: Microstructure, Properties, and Mater*; McGraw-Hill Education: Berkshire, UK, 2006; ISBN 0071589198.
2. Van Mier, J.G.M. *Concrete Fracture*; Taylor & Francis: New York, NY, USA, 2012; ISBN 978-1-4665-5470-2.
3. Qian, Z. Multiscale Modeling of Fracture Processes in Cementitious Materials. Ph.D. Thesis, Delft University of Technology, Delft, The Netherlands, 2012.
4. Constantinides, G.; Ulm, F.J. The effect of two types of C-S-H on the elasticity of cement-based materials: Results from nanoindentation and micromechanical modeling. *Cem. Concr. Res.* **2004**, *34*, 67–80. [CrossRef]
5. Hu, C.; Li, Z. A review on the mechanical properties of cement-based materials measured by nanoindentation. *Constr. Build. Mater.* **2015**, *90*, 80–90. [CrossRef]
6. Luo, Z.; Li, W.; Wang, K.; Shah, S.P. Research progress in advanced nanomechanical characterization of cement-based materials. *Cem. Concr. Compos.* **2018**, *94*, 277–295. [CrossRef]
7. Hu, C.; Li, Z. Micromechanical investigation of Portland cement paste. *Constr. Build. Mater.* **2014**, *71*, 44–52. [CrossRef]
8. Chen, S.J.; Duan, W.H.; Li, Z.J.; Sui, T.B. New approach for characterisation of mechanical properties of cement paste at micrometre scale. *Mater. Des.* **2015**, *87*, 992–995. [CrossRef]
9. Němeček, J.; Králík, V.; Šmilauer, V.; Polívka, L.; Jäger, A. Tensile strength of hydrated cement paste phases assessed by micro-bending tests and nanoindentation. *Cem. Concr. Compos.* **2016**, *73*, 164–173. [CrossRef]
10. Luković, M.; Schlangen, E.; Ye, G. Combined experimental and numerical study of fracture behaviour of cement paste at the microlevel. *Cem. Concr. Res.* **2015**, *73*, 123–135. [CrossRef]
11. Schlangen, E.; Lukovic, M.; Šavija, B.; Copuroglu, O. Nano-Indentation Testing and Modelling of Cement Paste. In *Concreep 10*; ASCE Library: New York, NY, USA, 2015; pp. 1028–1031. [CrossRef]
12. Zhang, H.; Šavija, B.; Figueiredo, S.C.; Lukovic, M.; Schlangen, E. Microscale testing and modelling of cement paste as basis for multi-scale modelling. *Materials* **2016**, *9*, 907. [CrossRef] [PubMed]
13. Zhang, H.; Šavija, B.; Schlangen, E. Combined experimental and numerical study on micro-cube indentation splitting test of cement paste. *Eng. Fract. Mech.* **2018**, *199*, 773–786. [CrossRef]
14. Zhang, H.; Šavija, B.; Figueiredo, S.C.; Schlangen, E. Experimentally validated multi-scale modelling scheme of deformation and fracture of cement paste. *Cem. Concr. Res.* **2017**, *102*, 175–186. [CrossRef]
15. Liu, D.; Flewitt, P.E.J. Deformation and fracture of carbonaceous materials using in situ micro-mechanical testing. *Carbon* **2017**, *114*, 261–274. [CrossRef]
16. Šavija, B.; Liu, D.; Smith, G.; Hallam, K.R.; Schlangen, E.; Flewitt, P.E.J. Experimentally informed multi-scale modelling of mechanical properties of quasi-brittle nuclear graphite. *Eng. Fract. Mech.* **2016**, *153*, 360–377. [CrossRef]
17. Lee, M.K.; Barr, B.I.G. An overview of the fatigue behaviour of plain and fibre reinforced concrete. *Cem. Concr. Compos.* **2004**, *26*, 299–305. [CrossRef]
18. Gao, L.; Hsu, C.T.T. Fatigue of concrete under uniaxial compression cyclic loading. *ACI Mater. J.* **1998**, *95*, 575–581.
19. Jiang, L.; Li, C.; Zhu, C.; Song, Z.; Chu, H. The effect of tensile fatigue on chloride ion diffusion in concrete. *Constr. Build. Mater.* **2017**, *151*, 119–126. [CrossRef]
20. Desmettre, C.; Charron, J.P. Water permeability of reinforced concrete subjected to cyclic tensile loading. *ACI Mater. J.* **2013**, *110*, 79–88. [CrossRef]
21. Van Breugel, K.; Koleva, D.; van Beek, T. *The Ageing of Materials and Structures: Towards Scientific Solutions for the Ageing of Our Assets*; Springer: New York, NY, USA, 2017; ISBN 9783319701943.
22. Hordijk, D. Local Approach to Fatigue of Concrete. Ph.D. Thesis, Delft University of Technology, Delft, The Netherlands, 1991.
23. Hsu, T.T.C. Fatigue and microcracking of concrete. *Matér. Constr.* **1984**, *17*, 51–54. [CrossRef]
24. Horii, H.; Shin, H.C.; Pallewatta, T.M. Mechanism of fatigue crack growth in concrete. *Cem. Concr. Compos.* **1992**, *14*, 83–89. [CrossRef]
25. Bažant, Z.P.; Xu, K. Size Effect in Fatigue Fracture of Concrete. *ACI Mater. J.* **1991**, *88*, 390–399.
26. Simon, K.M.; Chandra Kishen, J.M. A multiscale approach for modeling fatigue crack growth in concrete. *Int. J. Fatigue* **2017**, *98*, 1–13. [CrossRef]

27. Guo, L.P.; Carpinteri, A.; Roncella, R.; Spagnoli, A.; Sun, W.; Vantadori, S. Fatigue damage of high performance concrete through a 2D mesoscopic lattice model. *Comput. Mater. Sci.* **2009**, *44*, 1098–1106. [CrossRef]

28. Guo, L.P.; Carpinteri, A.; Spagnoli, A.; Sun, W. Experimental and numerical investigations on fatigue damage propagation and life prediction of high-performance concrete containing reactive mineral admixtures. *Int. J. Fatigue* **2010**, *32*, 227–237. [CrossRef]

29. Zhang, J.; Scherer, G.W. Comparison of methods for arresting hydration of cement. *Cem. Concr. Res.* **2011**, *41*, 1024–1036. [CrossRef]

30. Oliver, W.C.; Pharr, G.M. An improved technique for determining hardness and elastic modulus (Young's modulus). *J. Mater. Res.* **1992**, *7*, 1564–1583. [CrossRef]

31. Li, X.; Bhushan, B. A review of nanoindentation continuous stiffness measurement technique and its applications. *Mater. Charact.* **2002**, *48*, 11–36. [CrossRef]

32. Bouzakis, K.D.; Michailidis, N.; Hadjiyiannis, S.; Skordaris, G.; Erkens, G. The effect of specimen roughness and indenter tip geometry on the determination accuracy of thin hard coatings stress-strain laws by nanoindentation. *Mater. Charact.* **2002**, *49*, 149–156. [CrossRef]

33. Fischer-Cripps, A.C. *Factors Affecting Nanoindentation Test Data*; Springer: New York, NY, USA, 2000; ISBN 978-0-387-98914-3.

34. Poelma, R.H.; Morana, B.; Vollebregt, S.; Schlangen, E.; Van Zeijl, H.W.; Fan, X.; Zhang, G.Q. Tailoring the mechanical properties of high-aspect-ratio carbon nanotube arrays using amorphous silicon carbide coatings. *Adv. Funct. Mater.* **2014**, *24*, 5737–5744. [CrossRef]

35. Timoshenko, S. *Strength of Materials (Part I)*, 3rd ed.; D. Van Nostrand Co., Inc.: New York, NY, USA, 1958.

36. Stephens, L.S.; Kelly, K.W.; Simhadri, S.; McCandless, A.B.; Meletis, E.I. Mechanical property evaluation and failure analysis of cantilevered LIGA nickel microposts. *J. Microelectromech. Syst.* **2001**, *10*, 347–359. [CrossRef]

37. Patel, R.; Dubey, S.K.; Pathak, K.K. Effect of depth span ratio on the behaviour of beams. *Int. J. Adv. Struct. Eng.* **2014**, *6*, 3. [CrossRef]

38. Hutchinson, J.R. Shear Coefficients for Timoshenko Beam Theory. *J. Appl. Mech.* **2001**, *68*, 87–92. [CrossRef]

39. Stephen, N.G. Stephen2002—On a check on the accuracy of Timoshenko's beam theory.pdf. *J. Sound Vib.* **2002**, *257*, 809–812. [CrossRef]

40. Cowper, G.R. The Shear Coefficient in Timoshenko's Beam Theory. *J. Appl. Mech.* **1966**, *33*, 335–340. [CrossRef]

41. Bažant, Z.; Planas, J. *Fracture and Size Effect in Concrete and Other Quasibrittle Materiales*; CRC Press: Boca Raton, FL, USA, 1998; Volume 16, ISBN 084938284X.

42. Bažant, Z.P. Size Effect in Blunt Fracture: Concrete, Rock, Metal. *J. Eng. Mech.* **1984**, *110*, 518–535. [CrossRef]

43. Zhang, H.; Šavija, B.; Xu, Y.; Schlangen, E. Size effect on splitting strength of hardened cement paste: Experimental and numerical study. *Cem. Concr. Compos.* **2018**, *94*, 264–276. [CrossRef]

44. Zhang, B. Relationship between pore structure and mechanical properties of ordinary concrete under bending fatigue. *Cem. Concr. Res.* **1998**, *28*, 699–711. [CrossRef]

45. Thun, H.; Ohlsson, U.; Elfgren, L. Tensile fatigue capacity of concrete. *Nord. Concr. Res.* **2007**, *36*, 48–64.

46. Vicente, M.A.; González, D.C.; Mínguez, J.; Tarifa, M.A.; Ruiz, G.; Hindi, R. Influence of the pore morphology of high strength concrete on its fatigue life. *Int. J. Fatigue* **2018**, *112*, 106–116. [CrossRef]

47. Simon, K.M.; Kishen, J.M.C. Influence of aggregate bridging on the fatigue behavior of concrete. *Int. J. Fatigue* **2016**, *90*, 200–209. [CrossRef]

48. Holmen, J.O. Fatigue of Concrete by Constant and Variable Amplitude Loading. *ACI Spec. Publ.* **1982**, *75*, 71–110. [CrossRef]

49. Malek, A.; Scott, A.; Pampanin, S.; MacRae, G.; Marx, S. Residual Capacity and Permeability-Based Damage Assessment of Concrete under Low-Cycle Fatigue. *J. Mater. Civ. Eng.* **2018**, *30*, 4018081. [CrossRef]

50. Xu, B.X.; Yue, Z.F.; Wang, J. Indentation fatigue behaviour of polycrystalline copper. *Mech. Mater.* **2007**, *39*, 1066–1080. [CrossRef]

51. Saito, M. Characteristics of microcracking in concrete under static and repeated tensile loading. *Cem. Concr. Res.* **1987**, *17*, 211–218. [CrossRef]

52. Zhang, H.; Šavija, B.; Schlangen, E. Towards understanding stochastic fracture performance of cement paste at micro length scale based on numerical simulation. *Constr. Build. Mater.* **2018**, *183*, 189–201. [CrossRef]

53. Van Breugel, K. Simulation of Hydration and Formation of Structure in Hardening Cement-Based Materials. Ph.D. Thesis, Delft University of Technology, Delft, The Netherlands, 1991.

micromachines

MDPI

Article

Localized Deformation and Fracture Behaviors in InP Single Crystals by Indentation

Yi-Jui Chiu [1], Sheng-Rui Jian [2,*], Ti-Ju Liu [2], Phuoc Huu Le [3,*] and Jenh-Yih Juang [4,*]

[1]   School of Mechanical and Automotive Engineering, Xiamen University of Technology, No.600 Ligong Road, Jimei District, Xiamen 361024, China; chiuyijui@xmut.edu.cn
[2]   Department of Materials Science and Engineering, I-Shou University, Kaohsiung 840, Taiwan; diru@isu.edu.tw
[3]   Department of Physics and Biophysics, Faculty of Basic Sciences, Can Tho University of Medicine and Pharmacy, 179 Nguyen Van Cu Street, Can Tho 94000, Vietnam
[4]   Department of Electrophysics, National Chiao Tung University, Hsinchu 300, Taiwan
*   Correspondence: srjian@gmail.com (S.-R.J.); lhuuphuoc@ctump.edu.vn (P.H.L.); jyjuang@g2.nctu.edu.tw (J.-Y.J.); Tel.: +886-7-657-7711 (ext. 3130) (S.-R.J.)

Received: 11 October 2018; Accepted: 18 November 2018; Published: 22 November 2018

check for updates

Abstract: The indentation-induced deformation mechanisms in InP(100) single crystals were investigated by using nanoindentation and cross-sectional transmission electron microscopy (XTEM) techniques. The results indicated that there were multiple "pop-in" events randomly distributed in the loading curves, which were conceived to arise primarily from the dislocation nucleation and propagation activities. An energetic estimation on the number of nanoindentation-induced dislocations associated with pop-in effects is discussed. Furthermore, the fracture patterns were performed by Vickers indentation. The fracture toughness and the fracture energy of InP(100) single crystals were calculated to be around 1.2 MPa·m$^{1/2}$ and 14.1 J/m^2, respectively.

Keywords: InP(100) single crystal; Pop-in; nanoindentation; transmission electron microscopy; fracture toughness

1. Introduction

Nowadays, nanoindentation is extensively used to characterize the mechanical properties (such as hardness and elastic modulus) and elastic or plastic deformation behaviors of various nanoscale materials [1–6] and thin films [7–12]. In general, from the nanoindentation responses manifested in the load-displacement (*P-h*) curves, one can obtain the primary mechanical characteristics of the materials being measured. For instance, the onset of plastic deformation behaviors in crystalline materials is often characterized by sudden bursts of displacements at a nearly constant indentation load in the *P-h* curves. These phenomena, known as "pop-in," have been ubiquitously observed and often considered to be a result of dislocation activity during the nanoindentation process [13,14]. Lorenz et al. [15] proposed that the pop-in event is originated from homogeneous dislocation nucleation beneath the indenter tip. This scenario is reasonably in line with the low probability of encountering the pre-existing dislocations. Furthermore, previous cross-sectional transmission electron microscopy (XTEM) observations [14,16,17] evidenced the intimate correlations between pop-in events and dislocation activities in many nanoindentation studies [14–17].

Owing to its high electron velocity and the direct bandgap (~1.35 eV), zincblende-structured indium phosphide (InP) has been regarded as one of the most important III–V semiconductors. Currently, applications based on InP have been realized in a wide variety of electronic and photonic devices and systems, such as high-power and high-frequency electronics, solar cells, photodiodes, photodetectors, light-emitting diodes (LEDs), field effect transistors (FETs), and micro-electro-mechanical systems (MEMS) [18–20]. From an application point of view, in addition to its optoelectronic properties, a full understanding of the mechanical properties of InP is equally essential in order to widen its applications and manipulate the performance of devices.

For InP(100) single crystals, however, two different types of *P-h* curves obtained from nanoindentation studies were reported, namely single-discontinuity [21] and multi-discontinuities [22]. Such behavior remains intangible because the onset of nanoscale plasticity can be strongly influenced by various factors, such as indenter tip radius, temperature, and crystal plane [23,24]. Therefore, in order to understand nanoindentation-induced pop-in mechanisms and clarify the outstanding issues of nanoscale plasticity in InP(100) single crystals, a combination of XTEM and selected area diffraction (SAD) analyses were carried out in this work. The number of nanoindentation-induced dislocation loops in InP(100) single crystals was estimated within the context of the classical dislocation theory [25]. Moreover, the Vickers-indentation induced fracture toughness and fracture energy of InP (100) single crystals were calculated and discussed in details.

2. Materials and Methods

The (100)-oriented single-crystal InP used in this work was purchased from Semiconductor Wafer Inc. (Hsinchu, Taiwan) The nanoindentation tests were performed using a Nanoindenter MTS NanoXP® system (MTS Cooperation, Nano Instruments Innovation Center, Oak Ridge, TN, USA) with a diamond pyramid-shaped Berkovich indenter tip having a radius of curvature of ~50 nm. The mechanical properties of single-crystal InP(100) were obtained using the continuous stiffness measurements (CSM) technique, commonly practiced in the nanoindentation community [26]. Hardness and Young's modulus of single-crystal InP(100) were obtained using Oliver and Pharr method [27], as shown in Figure 1.

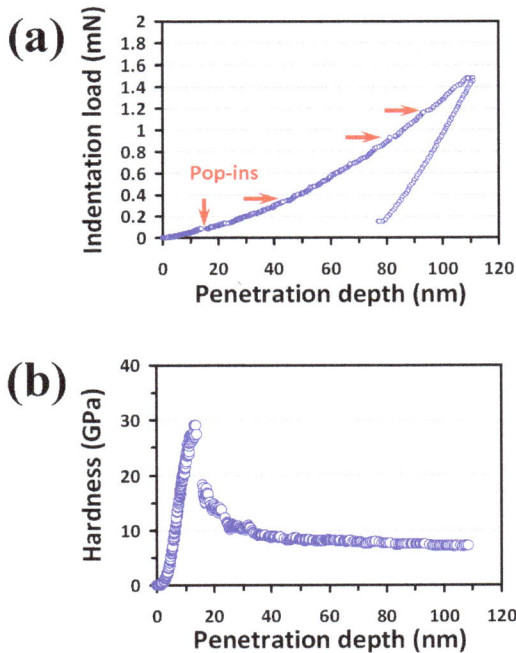

Figure 1. *Cont.*

(c)

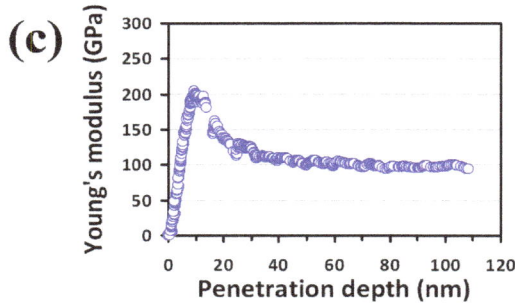

Figure 1. Nanoindentation results of single-crystal InP(100): (**a**) load-displacement curve showing the multiple "pop-ins" (arrows) during loading, (**b**) hardness-displacement curve and, (**c**) Young's modulus-displacement curve.

The hardness of the measured material is defined as the applied indentation loading divided by the projected contact area, $H = P_m/A_p$, where A_p is the projected contact area and P_m is the maximum indentation load. For a perfect Berkovich indenter, the projected area is given by $A_p = 24.56h_c^2$ with h_c being the contact depth. In addition, the elastic modulus of the material can be calculated based on the relationship proposed by Sneddon [28]: $S = 2\beta E_r \sqrt{A_p}/\sqrt{\pi}$. Here, S is the contact stiffness of the material and β is a geometric constant, with $\beta = 1.00$ for the Berkovich indenter. The reduced elastic modulus, E_r, can be calculated from the following equation:

$$\frac{1}{E_r} = \left(\frac{1-v^2}{E}\right)_d + \left(\frac{1-v^2}{E}\right)_{InP} \tag{1}$$

with v and E being Poisson's ratio and Young's modulus, respectively. The subscripts "*d*" and "*InP*" indicate the properties of the indenter and material, respectively. For the diamond indenter tip, $E_d = 1141$ GPa, $v_d = 0.07$ [27], and $v_{InP} = 0.25$ were assumed for single-crystal InP(100).

After being deformed by an indentation load of 150 mN, XTEM samples of InP(100) single crystals were prepared using a dual-beam focused ion beam (FIB) station (FEI Nova 220) with the lift-out technique. Pictorial illustrations of the FIB milling procedures are displayed in Figure 2. The sample was then picked up by a carbon membrane and placed on the TEM grid using a sharp glass tip under an optical microscopy (OM) outside the FIB station. The XTEM lamella was examined in a FEI TECNAI G^2 TEM operating at 200 kV.

Vickers indentation testing was carried out in single-crystal InP(100) to characterize the cracking behavior with five indents at a loading of 1.96 N by using a hardness tester (Akashi MVK-H11, Kanagawa, Japan). All indentations were made in ambient air at room temperature with a relative humidity of about 55%. The cracking patterns were examined and analyzed with an optical microscope (OM).

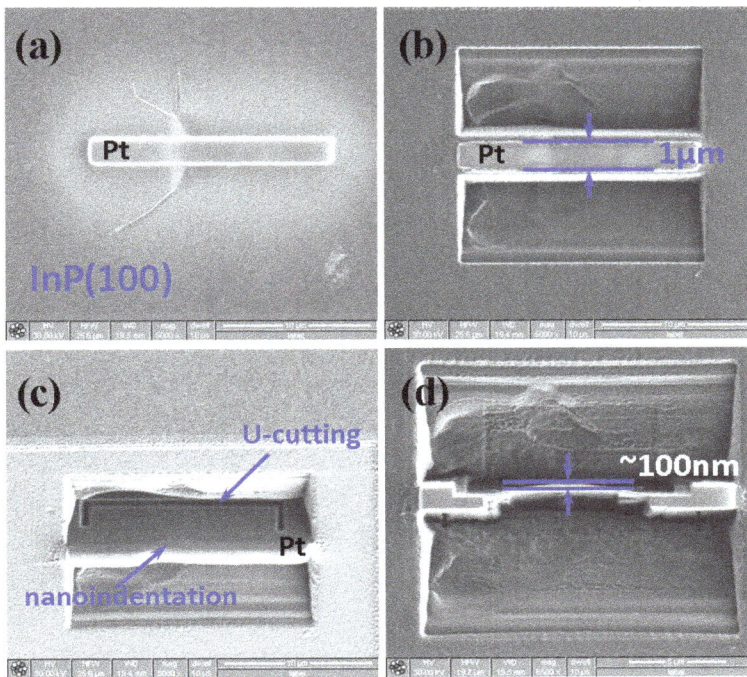

Figure 2. A typical procedure of focused ion beam (FIB) milling for single-crystal InP(100) is shown. Sample preparation starts with a line of nanoindentations. After depositing a protection layer of Pt (**a**), two big trenches are etched on either side of the indentation line by a high current ion beam (7~20 nA) (**b**) Further, the middle strip is thinned (**c**). An ion dose of 50 pA is used for final clearing steps and, finally thinned to a thickness of ~100 nm (**d**).

3. Results

3.1. Nanoindentation Responses

A typical CSM *P-h* curve of single-crystal InP(100) reflecting the elastic behavior and plastic deformation during nanoindentation is shown in Figure 1a. The results clearly show that there are several pop-ins occurring at different loading stages, as indicated by the arrows located at different indentation loadings, which is consistent with a previous report [22]. Hardness and Young's modulus versus penetration depth curves obtained from the CSM analyses for single-crystal InP(100) are displayed in Figure 1b,c, respectively. Both curves exhibit very similar depth-dependent trends, namely an initial quasi-linear increase to a maximum value within the first 10–15 nm, followed by a subsequent steep decrease in the 20–30-nm range, and finally reaching a constant value. It is interesting to note that the steep decrease after the first stage essentially coincides with where the first pop-in event is observed, indicating that a bursting activity of dislocation might have occurred.

For a uniform material, hardness and Young's modulus do not change significantly with increasing penetration depth. The initial increase seen in Figure 1b,c is owing to the fact that the practical indenter tip is of a finite radius of shape point. This effectively sets the limit on the indentation depth that is necessary to obtain reliable hardness and Young's modulus records of the measured material. As shown in Figure 1b,c, hardness and Young's modulus reach a constant value at a similar moderate indentation depth. Thus, the values of hardness and Young's modulus obtained at this stage can be regarded as intrinsic properties of single-crystal InP(100). In this report, both mechanical parameters were determined by taking the average values within a penetration depth ranging from 60 nm to100 nm.

Hardness and Young's modulus of single-crystal InP(100) thus obtained are about 7.5 GPa and 101.8 GPa, respectively. These values are both substantially larger than those reported by Bradby et al. [21], where hardness and Young's modulus for InP(100) were ~5.1 GPa and ~82 GPa, respectively. We note that in their experiments, a spherical indenter with a radius of ~4.2 μm and a load up to 50 mN were used, whereas in the present study a pyramid shape Berkovich indenter with a tip radius of ~40 nm (facing 65.3° from the vertical axis) and a typical load of less than 2 mN were used. It is reasonable to speculate that in the present study the probed deformation region could be more localized, which might also give rise to the apparent discrepancies between the mechanical parameters obtained from different experimental set-ups and operation modes.

Within the dominant deformation mechanism in the context of dislocation, the multiple "pop-ins" (indicated by the arrows in Figure 1a) can be regarded as the trigger of sudden collective activities of dislocation [29–31] (such as dislocation generation or movement bursts), giving rise to the seemingly discontinuous plastic deformation during nanoindentation. Such massive dislocation activities are also consistent with the conjectures of the resultant "noisy" features seen in the depth-dependent curves of hardness and Young's modulus (Figure 1b,c), as well as those reported by Almeida et al. [32] and Jian et al. [22]. The multiple "pop-in" behaviors had also been observed in Reference [32]. From Figure 1a, the first "pop-in" is observed on the loading curve at a load of about 0.08 mN in the present work, which is substantially smaller than that (~0.2 mN) reported previously by Almeida et al. [32]. It is noted that the indenter used in Reference [32] was a cono-spherical-type tip with a radius of ~260 nm equipped in a Hysitron Triboscope nanoindenter system. It is possible that different operating modes, indenter geometrical shapes, and size of radius may lead to the vastly dissimilar nanoindentation results.

However, as mentioned above, when a spherical indenter with a larger tip radius was used, only a single "pop-in" event was observed [21]. This discrepancy, as we discussed above, might originate primarily from the differences in operating modes and the geometric shape of the indenter being used to probe the nanoindentation properties of the same material. The other feature to be noted is that no evidence of reverse discontinuities in the unloading segment (the so-called "pop-out") can be identified in the present case. This indicates that the pressure-induced phase transformation commonly observed in single-crystal silicon [33] is probably not happening in InP, although the zincblende crystalline structure of InP is in fact not that different from the diamond structure of Si. In any case, in order to gain a more comprehensive understanding of the underlying indentation-induced deformation mechanism, direct microstructural investigations such as SEM and XTEM analyses are certainly indispensable.

3.2. XTEM and SAD Analyses

Figure 3a shows a SEM image of the InP(100) single-crystal surface after being indented with a load of 150 mN, featuring the characteristics of Berkovich nanoindentation-induced cracks. Although it appears that the paths of cracking propagation are not straight and the propagation directions are somewhat random, the directions, nevertheless, can be roughly divided into <100> and <110> directions [22]. A bright-field XTEM image of the area immediately beneath the tip of the Berkovich indenter is shown in Figure 3b. It is clearly evident that the slip bands are oriented at ~54.7° to the (100) surface, indicating that the dislocations have been gliding along the <110>{111} slip systems expected for the zincblende-structured InP. The rosette arm patterns ubiquitously observed in materials with similar crystal structure [32,34] are also evident, as indicated in Figure 3b. A closer examination by select area diffraction (SAD) further reveals that the deformation zone immediately beneath the indent consists of a mixture of slip-bands and micro-twins. This is clearly illustrated by the double spots and streaks seen in the [011] zone-axis SAD pattern displayed in Figure 3c. The SAD pattern also indicates that the dislocations and twins are lying parallel on the {111} planes. The direct microstructural observations by XTEM, thus, clearly confirm that indentation-induced deformation in single-crystal InP(100) is exclusively dominated by dislocation activities, and mechanisms such as phase transformation and amorphization are not involved in the nanoindentation process. In the present study, the Berkovich nanoindentation-induced multiple "pop-in" behaviors exhibited on the loading curve are attributed to the formation of slip-bands and micro-twins in single-crystal InP(100). Whereas, based on their observations, Almeida et al. [32] proposed that the multiple "pop-in" behaviors resulted from the Lomer-Cottrell locks, work-hardened region, and a high density of dislocation loops formed beneath the cono-spherical tip. From this viewpoint, the geometrical shape of the indenter tip may have played an important role on the activation of the slip system or/and the formation of dislocations. This may also explain the different number of pop-in events often observed in materials with the same crystal structure during nanoindentation, such as zincblende-structured InP [21,22,32] and hexagonal-structured GaN thin films [14,16].

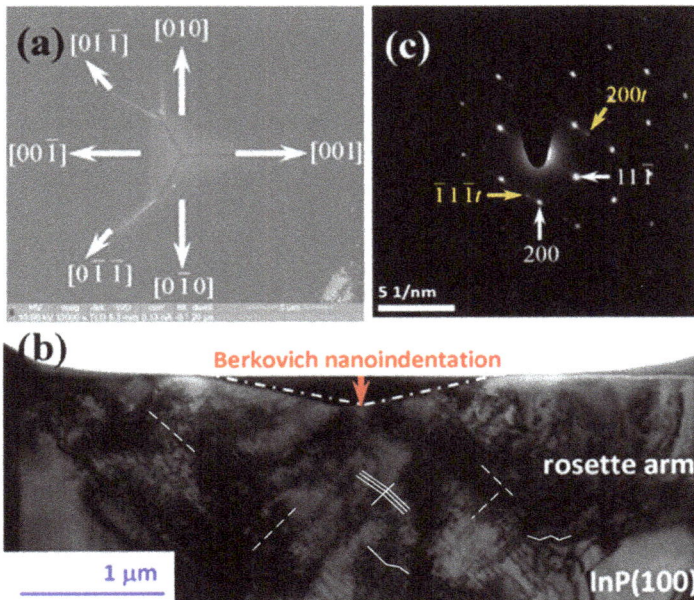

Figure 3. An indented InP(100) single crystal under an indentation load of 150 mN. (**a**) SEM micrograph showing the cracking behaviors. (**b**) Bright-field XTEM image: micro-twins are indicated by solid white lines; dash lines are used to guide the eyes for lattice fringes. (**c**) SAD pattern of sample underneath the Berkovich indenter.

Moreover, from Figure 3b, the number of dislocations (N) generated by the indenter can be roughly estimated using the following expression: $N = h_r/b_z$, where h_r is the residual depth of the indenter and b_z is the component of the dislocations along the loading axis [35]. Taking h_r ~1200 nm, it is estimated that there were about 3000 dislocations formed underneath the indenter tip in this work. This number is about an order of magnitude smaller than the dislocation loops with critical size generated during first pop-ins obtained using thermodynamics energetic estimation (see below). However, considering that the above estimation was made based on the TEM image taken after the entire indentation was completed and the applied stress was removed, this, in fact, is quite consistent. This is because with further increasing load, the dislocation loops formed during the first pop-in may slide and merge to form twins and slip-bands, resulting in multiple pop-ins in the later stage of indentation and reducing the number of dislocations in the residual indentation depth, as revealed in the TEM image. Further, the average dislocation densities (ρ) were estimated using the relationship

$\rho = 2N/Lt$ [36], where L is the total length of random lines projected on a given area of XTEM image and t is the foil thickness. In this experimental result, ρ is thus estimated in the order of about 10^{14} m^{-2}.

3.3. Homogeneous Dislocation Nucleation

In the scenario described above, the first pop-in event appearing in the loading segment naturally reflects the onset of plasticity in single-crystal InP(100) manifested by sudden dislocation nucleation and propagation. In other words, the corresponding loading is likely intimately associated with the critical shear stress (τ_{max}), and the energy associated with the pop-in depth may directly account for the number of indentation-induced newly nucleated dislocation loops. Following the analytical model by Johnson [37], τ_{max} can be related to an indentation load (P_c) at which a discontinuity in the load-displacement curve takes place, through the following equation:

$$\tau_{max} = 0.31 \left(\frac{6P_c E_r^2}{\pi^3 R^2} \right)^{1/3} \tag{2}$$

where R is the radius of the indenter tip. Thus, the obtained τ_{max} for single-crystal InP(100) is about 2.3 GPa. We assume that the τ_{max} is responsible for the homogeneous dislocation nucleation underneath the indenter tip. Therefore, the shear stress that initiates plastic deformation and the energy required for generating a dislocation loop to prevail the deformation can be estimated from the present data. The free energy (U) of a circular dislocation loop of radius (r) is given as:

$$U = \gamma_{dis} 2\pi r - \tau b \pi r^2 \tag{3}$$

where γ_{dis} is the line energy of the dislocation loop, b is the magnitude of Burgers vector ($\sim$0.4 nm) [38], and τ is the external shear stress acting on the dislocation loop. The energy required to create a dislocation loop in a defect-free lattice is described in the first term on the right-hand side of Equation (3) and is also equal to the increased lattice energy due to the formation of a dislocation loop. The second term of Equation (3) is nothing but the strain energy released via work done by the applied stress (τ) to expand the dislocation loop over a displacement of one Burgers vector. The lattice strain in the vicinity of the dislocation for $r > r_{core}$, thus γ_{dis}, is given by Reference [25]:

$$\gamma_{dis} = \frac{Gb^2}{8\pi} \frac{2 - v_{InP}}{1 - v_{InP}} \left[\ln \left(\frac{4r}{r_{core}} \right) - 2 \right] \tag{4}$$

where G is the shear modulus, and for single-crystal InP(100) $G \approx 31$ GPa [39]. The value of the radius of dislocation core r_{core} is usually assumed to be about one lattice constant. By using Equations (1), (3) and (4) can be rewritten as:

$$U = \frac{Gb^2}{4} \left(\frac{2 - v_{InP}}{1 - v_{InP}} \right) \left(\ln \frac{4r}{r_{core}} - 2 \right) r - \pi b r^2 \tau_c \tag{5}$$

This relates the material properties and observed pop-in load to the free energy responsible for dislocation nucleation. The resolved shear stress (τ_c) is usually taken as the half of τ_{max} [39]. The U has a maximum at a critical radius (r_c) above which the system gains energy by increasing r. According to Equation (5), this maximum energy decreases with increasing load and a pop-in, i.e., the homogeneous formation of a circular dislocation loop becomes possible without thermal energy at $U = 0$ [40]. With this condition and setting $dU/dr = 0$ for a maximum, this yields $\tau_c = 2\gamma_{dis}/br$ and $r_c = \left(e^3 r_{core} \right)/4$. Consequently, $r_{core} \approx 0.43$ nm and $r_c = 2.15$ nm are obtained. The value $r_{core} \approx 0.43$ nm is slightly smaller than $a \approx 0.587$ nm for InP, however, it is consistent with the atomic distance along the <110> orientation ($\approx$0.42 nm), indicating that the above analysis is reasonable. The fact that the critical radius of the dislocation loop r_c (=2.15 nm) needs to extend over a distance of about 5 times the dislocation core to become stable is also interesting. This suggests that the system is more prone to accommodate strain energy with larger dislocations loops.

The number of dislocation loops formed in the first pop-in can be estimated from the associated work (W_p) done during nanoindentation. As depicted in Figure 4, W_p is estimated to be about 0.2×10^{-12} Nm, implying that ~3×10^4 dislocation loops with a radius larger than r_c have been generated during the pop-in event. This number is relatively low and consistent with the scenario of homogeneous dislocation nucleation-induced pop-in, instead of activated collective motion of pre-existing dislocations [15]. Moreover, as discussed above, this number is also in line with the number of dislocations estimated from the TEM image taken from the region of residual indentation depth. Alternatively, one can take the total dissipation energy as the energy to estimate the number of dislocations with critical radius being generated during entire nanoindentation practice. In that case, as many as ~10^6 dislocation loops may have been formed during nanoindentation. However, this number can only be regarded as an upper limit because it is quite unlikely that all the dissipated indentation energy is completely transferred to form dislocation loops within the deformation region in single-crystal InP(100).

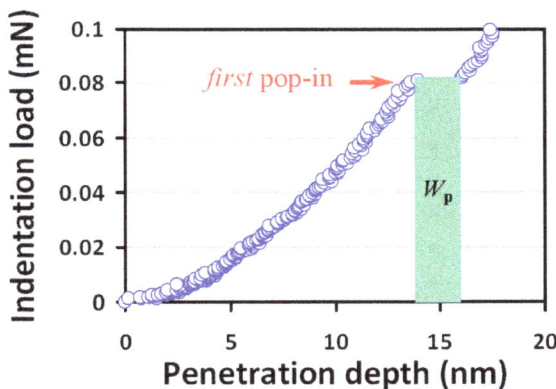

Figure 4. The corresponding pop-in event (red arrow) from Figure 2a is zoomed in. The plastic strain work is W_p: critical loading × sudden incremental displacement.

3.4. Vickers Indentation Induced Fracture Behavior

Fracture toughness (K_C) can be readily measured by the induced cracks at the corners of indentations made on the substrate materials. This method is known as the indentation microfracture [41–43]. Figure 5 shows the Vickers-indentation-induced cracking pattern on single-crystal InP(100) at a loading of 1.96 N. It can be found that the ratio of average cracking length ($l = (c_1 + c_2)/4$, where c_1 and c_2 are defined in Figure 5) to the half-diagonal of the indentation (a) meets the criteria of Palmqvist cracks with $0.25 \leq l/a \leq 2.5$. Therefore, the formula proposed by Niihara et al. [41,42] was adopted to calculate the K_C of single-crystal InP(100) here, as follows:

$$K_C = 0.009 \left(\frac{E_{InP}}{H} \right)^{2/5} \frac{P_a}{a\sqrt{l}} \tag{6}$$

where P_a is the applied load. The K_C of single-crystal InP(100) obtained is about 1.2 MPa·m$^{1/2}$. Comparing to the K_{IC} values of 0.42~0.53 MPa·m$^{1/2}$ for InP reported by Ericson et al. [44], the present result seems slightly too large. However, considering that K_C is strongly dependent on specimen geometry and usually decreases with increasing thickness to reach a minimum value known as K_{IC}, we believe that a factor of 2–2.5 difference is quite reasonable. Finally, the fracture energy (G_C) can also be calculated using the equation $G_C = K_C^2 \left[(1 - v_{InP}^2)/E_{InP} \right]$ [45], yielding approximately 14.1 J·m^{-2}

for single-crystal InP(100). This number indicates that InP is, in fact, quite ductile, which is also consistent with the dislocation-dominated deformation discussed above.

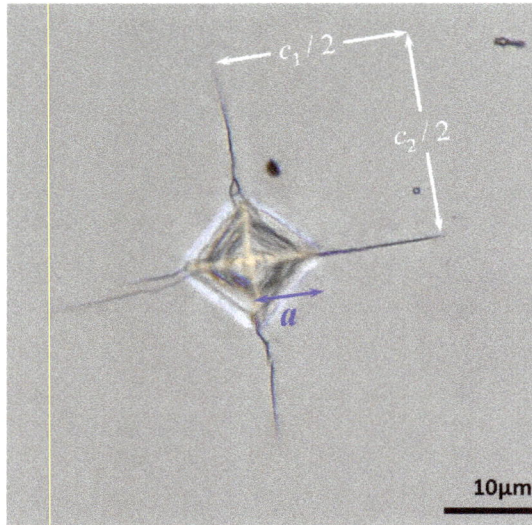

Figure 5. Vickers indentation at 1.96 N in single-crystal InP(100): Palmqvist cracks emitting from Vickers indentation, where a is the half-diagonal of the indentation and $l = (c_1 + c_2)/4$ is the average length of the radial cracks for each indentation.

4. Conclusions

To sum up, the nano- and micro-scale deformation mechanisms and behaviors of single-crystal InP(100) were studied by combining indentation, SEM, and XTEM techniques. From the XTEM results and SAD analysis, the nanoindentation responses of single-crystal InP(100) involve the the formation of slip-bands and micro-twins. Preliminary energetic estimations indicate that the number of dislocation loops induced by nanoindentation to trigger the plastic deformation accounting for the first pop-in event was in the order of 10^4 with a critical radius $r_c \approx 2.15$ nm. Furthermore, from Vickers indentation tests, the obtained values of K_C and G_C of single-crystal InP(100) were about 1.2 MPa·m$^{1/2}$ and 14.1 Jm^{-2}, respectively.

Author Contributions: Y.-J.C. and T.-J.L. contributed to the nanoindentation experiments and analyses. S.-R.J., J.-Y.J., and P.H.L. contributed to the discussion on materials characterizations. S.-R.J. designed the experiments and drafted the manuscript. All authors read and approved the final manuscript.

Funding: Financial support from the Ministry of Science and Technology, Taiwan under Contract Nos.: MOST 106-2112-M009-013-MY3, MOST 103-2112-M-009-015-MY3, MOST 107-2112-M-214-001, MOST 106-2112-M-214-001, and MOST 105-2112-M-214-001. This work was supported by the Fujian Nature foundation, No. 2016J01039; Xiamen City Project No. 3502Z20173037.

Conflicts of Interest: The authors declare no conflict of interest.

References

1. Li, X.D.; Gao, H.; Murphy, C.J.; Gou, L. Nanoindentation of Cu$_2$O Nanocubes. *Nano Lett.* **2004**, *4*, 1903–1907. [CrossRef]
2. Li, X.D.; Wang, X.; Xiong, Q.; Eklund, P.C. Mechanical Properties of ZnS Nanobelts. *Nano Lett.* **2005**, *5*, 1982–1986. [CrossRef] [PubMed]
3. Raichman, Y.; Kazakevich, M.; Rabkin, E.; Tsur, Y. Inter-Nanoparticle Bonds in Agglomerates Studied by Nanoindentation. *Adv. Mater.* **2006**, *18*, 2028–2030. [CrossRef]

4.   Tao, X.; Li, X.D. Catalyst-Free Synthesis, Structural, and Mechanical Characterization of Twinned Mg$_2$B$_2$O$_5$ Nanowires. *Nano Lett.* **2008**, *8*, 505–510. [CrossRef] [PubMed]

5.   Bao, L.; Xu, Z.H.; Li, R.; Li, X.D. Catalyst-Free Synthesis and Structural and Mechanical Characterization of Single Crystalline Ca$_2$B$_2$O$_5$·H$_2$O Nanobelts and Stacking Faulted Ca$_2$B$_2$O$_5$ Nanogrooves. *Nano Lett.* **2010**, *10*, 255–262. [CrossRef] [PubMed]

6.   Jian, S.R.; Sung, T.H.; Huang, J.C.; Juang, J.Y. Deformation behaviors of InP pillars under uniaxial compression. *Appl. Phys. Lett.* **2012**, *101*, 151905. [CrossRef]

7.   Jian, S.R.; Chen, G.J.; Lee, J.W. Effects of annealing temperature on nanomechanical and microstructural properties of Cu-doped In$_2$O$_3$ thin films. *Appl. Phys. A* **2017**, *123*, 726. [CrossRef]

8.   Jian, S.R.; Le, P.H.; Luo, C.W.; Juang, J.Y. Nanomechanical and wettability properties of Bi$_2$Te$_3$ thin films: Effects of post-annealing. *J. Appl. Phys.* **2017**, *121*, 175302. [CrossRef]

9.   Lai, H.D.; Jian, S.R.; Tuyen, L.T.C.; Le, P.H.; Luo, C.W.; Juang, J.Y. Nanoindentation of Bi2Se3 Thin Films. *Micromachines* **2018**, *9*, 518. [CrossRef] [PubMed]

10. Chen, G.J.; Jian, S.R. Effects of Cu doping on the structural and nanomechanical properties of ZnO thin films. *Appl. Phys. A* **2018**, *124*, 575. [CrossRef]

11. Chiu, Y.J.; Shen, C.Y.; Chang, H.W.; Jian, S.R. Characteristics of Iron-Palladium alloy thin films deposited by magnetron sputtering. *Results Phys.* **2018**, *9*, 17–22. [CrossRef]

12. Chen, G.J.; Jian, S.R.; Juang, J.Y. Surface Analysis and Optical Properties of Cu-Doped ZnO Thin Films Deposited by Radio Frequency Magnetron Sputtering. *Coatings* **2018**, *8*, 266. [CrossRef]

13. Morris, J.R.; Bei, H.; Pharr, G.M.; George, E.P. Size Effects and Stochastic Behavior of Nanoindentation Pop In. *Phys. Rev. Lett.* **2011**, *106*, 165502. [CrossRef] [PubMed]

14. Chien, C.H.; Jian, S.R.; Wang, C.T.; Juang, J.Y.; Huang, J.C.; Lai, Y.S. Cross-sectional transmission electron microscopy observations on the Berkovich indentation-induced deformation microstructures in GaN thin films. *J. Phys. D Appl. Phys.* **2007**, *40*, 3985–3990. [CrossRef]

15. Lorenz, D.; Zeckzer, A.; Hilpert, U.; Grau, P.; Johnson, H.; Leipner, H.S. Pop-in effect as homogeneous nucleation of dislocations during nanoindentation. *Phys. Rev. B* **2003**, *67*, 172101. [CrossRef]

16. Bradby, J.E.; Kucheyev, S.O.; Williams, J.S.; Leung, J.W.; Swain, M.V.; Munroe, P.; Li, G.; Phillips, M.R. Indentation-induced damage in GaN epilayers. *Appl. Phys. Lett.* **2002**, *80*, 383. [CrossRef]

17. Bradby, J.E.; Williams, J.S.; Leung, J.W.; Swain, M.V.; Munroe, P. Nanoindentation-induced deformation of Ge. *Appl. Phys. Lett.* **2002**, *80*, 2651. [CrossRef]

18. Robidas, D.; Arunseshan, C.; Deepthi, K.R.; Arivuoli, D. Nanomechanical characterization of indium phosphide epilayer using nanoindentation technique. *Int. J. Mech. Ind. Eng.* **2013**, *3*, 22–26.

19. Zafar, F.; Iqbal, A. Indium phosphide nanowires and their applications in optoelectronic devices. *Proc. R. Soc. A* **2016**, *472*, 20150804. [CrossRef] [PubMed]

20. Tay, C.J.; Quan, C.; Gopal, M.; Shen, L.; Akkipeddi, R. Nanoindentation techniques in the measurement of mechanical properties of InP-based free-standing MEMS structures. *J. Micromech. Microeng.* **2008**, *18*, 025015. [CrossRef]

21. Bradby, J.E.; Williams, J.S.; Leung, J.W.; Swain, M.V.; Munroe, P. Mechanical deformation of InP and GaAs by spherical indentation. *Appl. Phys. Lett.* **2001**, *78*, 3235. [CrossRef]

22. Jian, S.R.; Jang, J.S.C. Berkovich nanoindentation on InP. *J. Alloys Compd.* **2009**, *482*, 498–501. [CrossRef]

23. Lu, J.Y.; Ren, H.; Deng, D.M.; Wang, Y.; Chen, K.J.; Lau, K.M.; Zhang, T.Y. Thermally activated pop-in and indentation size effects in GaN films. *J. Phys. D Appl. Phys.* **2012**, *45*, 085301. [CrossRef]

24. Huang, J.; Xu, K.; Fan, Y.M.; Niu, M.T.; Zeng, X.H.; Wang, J.F.; Yang, H. Nanoscale anisotropic plastic deformation in single crystal GaN. *Nanoscale Res. Lett.* **2012**, *7*, 150. [CrossRef] [PubMed]

25. Hirth, J.P.; Lothe, J. *Theory of Dislocations*, 2nd ed.; John Wiley and Sons: Hoboken, NJ, USA, 1982.

26. Li, X.D.; Bhushan, B. A review of nanoindentation continuous stiffness measurement technique and its applications. *Mater. Charact.* **2002**, *48*, 11–36. [CrossRef]

27. Oliver, W.C.; Pharr, G.M. An improved technique for determining hardness and elastic modulus using load and displacement sensing indentation experiments. *J. Mater. Res.* **1992**, *7*, 1564–1583. [CrossRef]

28. Sneddon, I.N. The relation between load and penetration in the axisymmetric Boussinesq problem for a punch of arbitrary profile. *Int. J. Eng. Sci.* **1965**, *3*, 47–57. [CrossRef]

29. Jian, S.R.; Ku, S.A.; Luo, C.W.; Juang, J.Y. Nanoindentation of GaSe thin films. *Nanoscale Res. Lett.* **2012**, *7*, 403. [CrossRef] [PubMed]

30. Jian, S.R.; Tasi, C.H.; Huang, S.Y.; Luo, C.W. Nanoindentation pop-in effects of Bi2Te3 thermoelectric thin films. *J. Alloys Compd.* **2015**, *622*, 601–605. [CrossRef]

31. Mosca, D.H.; Mattoso, N.; Lepienski, C.M.; Veiga, W.; Mazzaro, I.; Etgens, V.H.; Eddrief, M. Mechanical properties of layered InSe and GaSe single crystals. *J. Appl. Phys.* **2002**, *91*, 140. [CrossRef]

32. Almeida, C.M.; Prioli, R.; Wei, Q.Y.; Ponce, F.A. Early stages of mechanical deformation in indium phosphide with the zinc blende structure. *J. Appl. Phys.* **2012**, *112*, 063514. [CrossRef]

33. Jian, S.R.; Chen, G.J.; Juang, J.Y. Nanoindentation-induced phase transformation in (1 1 0)-oriented Si single-crystals. *Curr. Opin. Solid State Mater. Sci.* **2010**, *14*, 69–74. [CrossRef]

34. Patriarche, G.; Le Bourhis, E. Low-load deformation of InP under contact loading; comparison with GaAs. *Phil. Mag. A* **2002**, *82*, 1953–1961. [CrossRef]

35. Le Bourhis, E.; Patriarche, G. TEM-nanoindentation studies of semiconducting structures. *Micron* **2007**, *38*, 377–389. [CrossRef] [PubMed]

36. Ham, R.K. The determination of dislocation densities in thin films. *Phil. Mag.* **1961**, *6*, 1183–1184. [CrossRef]

37. Johnson, K.L. *Contact Mechanics*; Cambridge University Press: Cambridge, UK, 1985.

38. Yonenaga, I.; Suzuki, T. Indentation hardnesses of semiconductors and a scaling rule. *Phil. Mag. Lett.* **2002**, *82*, 535–542. [CrossRef]

39. Chiu, Y.L.; Ngan, A.H.W. Time-dependent characteristics of incipient plasticity in nanoindentation of a Ni3Al single crystal. *Acta Mater.* **2002**, *50*, 1599–1611. [CrossRef]

40. Leipner, H.S.; Lorenz, D.; Zeckzer, A.; Lei, H.; Grau, P. Nanoindentation pop-in effect in semiconductors. *Phys. B* **2001**, *308–310*, 446–449. [CrossRef]

41. Niihara, K.; Morena, R.; Hasselman, D.P.H. Evaluation of K_{Ic} of brittle solids by the indentation method with low crack-to-indent ratios. *J. Mater. Sci. Lett.* **1982**, *1*, 13–16. [CrossRef]

42. Niihara, K.; Morena, R.; Hasselman, D.P.H. A fracture mechanics analysis of indentation-induced Palmqvist crack in ceramics. *J. Mater. Sci. Lett.* **1983**, *2*, 221–223. [CrossRef]

43. Schiffmann, K.I. Determination of fracture toughness of bulk materials and thin films by nanoindentation: Comparison of different models. *Philos. Mag.* **2011**, *91*, 1163–1178. [CrossRef]

44. Ericson, F.; Johansson, S.; Schweitz, J.-A. Hardness and fracture toughness of semiconducting materials studied by indentation and erosion techniques. *Mater. Sci. Eng. A* **1988**, *105–106*, 131–141. [CrossRef]

45. Rafiee, M.A.; Rafiee, J.; Srivastava, I.; Wang, Z.; Song, H.; Yu, Z.Z.; Koratkar, N. Fracture and Fatigue in Graphene Nanocomposites. *Small* **2010**, *6*, 179–183. [CrossRef] [PubMed]

micromachines

MDPI

Article

Calibration of a Constitutive Model from Tension and Nanoindentation for Lead-Free Solder

Xu Long [1,*], Xiaodi Zhang [2], Wenbin Tang [3], Shaobin Wang [4], Yihui Feng [5] and Chao Chang [6]

[1] School of Mechanics and Civil & Architecture, Northwestern Polytechnical University, Xi'an 710072, China
[2] College of Mining Engineering, Liaoning Shihua University, Fushun 113001, China; zhangxiaodiw@163.com
[3] School of Mechanics and Civil & Architecture, Northwestern Polytechnical University, Xi'an 710072, China; tangwb@mail.nwpu.edu.cn
[4] School of Mechanics and Civil & Architecture, Northwestern Polytechnical University, Xi'an 710072, China; shaobinwang@mail.nwpu.edu.cn
[5] State Key Laboratory of Nonlinear Mechanics, Institute of Mechanics, Chinese Academy of Sciences, Beijing 100190, China; fengyh@lnm.imech.ac.cn
[6] School of Applied Science, Taiyuan University of Science and Technology, Taiyuan 030024, China; cc@tyust.edu.cn
*   Correspondence: xulong@nwpu.edu.cn; Tel.: +86-029-88431000

Received: 5 October 2018; Accepted: 14 November 2018; Published: 20 November 2018

check for
updates

Abstract: It is challenging to evaluate constitutive behaviour by using conventional uniaxial tests for materials with limited sizes, considering the miniaturization trend of integrated circuits in electronic devices. An instrumented nanoindentation approach is appealing to obtain local properties as the function of penetration depth. In this paper, both conventional tensile and nanoindentation experiments are performed on samples of a lead-free Sn–3.0Ag–0.5Cu (SAC305) solder alloy. In order to align the material behaviour, thermal treatments were performed at different temperatures and durations for all specimens, for both tensile experiments and nanoindentation experiments. Based on the self-similarity of the used Berkovich indenter, a power-law model is adopted to describe the stress–strain relationship by means of analytical dimensionless analysis on the applied load-penetration depth responses from nanoindentation experiments. In light of the significant difference of applied strain rates in the tensile and nanoindentation experiments, two "rate factors" are proposed by multiplying the representative stress and stress exponent in the adopted analytical model, and the corresponding values are determined for the best predictions of nanoindentation responses in the form of an applied load–indentation depth relationship. Eventually, good agreement is achieved when comparing the stress–strain responses measured from tensile experiments and estimated from the applied load–indentation depth responses of nanoindentation experiments. The rate factors ψ_σ and ψ_n are calibrated to be about 0.52 and 0.10, respectively, which facilitate the conversion of constitutive behaviour from nanoindentation experiments for material sample with a limited size.

Keywords: nanoindentation; constitutive model; rate factor; dimensionless analysis; solder

1. Introduction

As described by the observation of Moore's law, the miniaturization of electronic devices is still continuously ongoing, despite material and manufacturing challenges [1]. It is difficult to obtain the constitutive behaviour of new emerging electronic packaging materials at such a local scale. Even though conventional tensile experiments can be performed to obtain the stress–strain relationship, the specimens have to be designed with a sufficient size to be conveniently clamped. The requirement of specimen geometry is difficult to meet for some die-attach pastes, such as silver nanoparticle paste, with the evaporation during the sintering process. The microstructure and also the material properties of sintered material in a great volume is distinguishingly different from the die-attach form in actual applications, due to the coalescence of sintered nanoparticles [2]. The material properties of lead-containing solders have been well-known and applied with an outstanding mechanical reliability; nevertheless, these solders have been replaced by lead-free solder alloys in recent years around the world. Compared with lead-containing solders, the lead-free alloys' capacity to resist thermo-mechanical fatigue and electromigration is better, but is also detrimental to mechanical shock and whisker growth [3]. As reviewed by Zhang and Tu [4], extensive studies on the composite Pb-free solders by adding nanoparticles has been conducted to strengthen the physical and solder mechanical properties, such as wettability, creep resistance, and hardness. Xu et al. [5] demonstrated that the Sn–Ag–Cu solder paste added by FeCo magnetic nanoparticles can be reflowed locally with alternating current magnetic fields, so that the interconnects form in area array packages, and the eddy current heating in the printed circuit board is minimized.

In light of the miniaturization trend of electronic devices, it is challenging to obtain the constitutive behaviour by using conventional uniaxial tests for small-sized material samples, as adopted during packaging applications. Instrumented nanoindentation is a suitable approach to quantify the material properties at a smaller scale, which is feasible for the die-attach materials, especially with a limited size of electronic devices. Long et al. [6] determined the strain-rate sensitivity of several types of die-attached materials, using nanoindentation with multiple strain-rate jumps accompanied by a continuous stiffness measurement. Zhang et al. [7] investigated the size effect of surface pit defects on the yield load of thin film, by using the quasi-continuum method to simulate nanoindentation. Bo et al. [8] determined the mechanical properties of cells using the stress–relaxation curve from the indentation process with an atomic force microscope. Recently, Rengel et al. [9] measured the mechanical behaviour of a hydride blister to reduce the mechanical and fracture properties of nuclear fuel cladding. Lee et al. [10] performed nanoindentation under the temperature range between 25 °C and 300 °C to determine the activation energy for the plastic flow in a nanocrystalline CoCrFeMnNi high-entropy alloy. Chu, et al. [11] investigated the mechanical properties of Fe–Zr welded joints, as well as dependence with the microstructures using nanoindentation. Hsueh et al. [12] revealed the size effect and strain-induced double twin in duplex stainless steel, prepared using the activated gas tungsten arc welding technique. Using both indentation and tensile tests for fully dense nanocrystalline nickel, Schwaiger et al. [13] found that the strain-rate sensitivity of deformation is strongly related to the grain size. Phani and Oliver [14] demonstrated that uniaxial creep behaviour over a wide range of strain rates and temperatures agrees well with the uniaxial creep behavior using high temperature nanoindentation. Humphrey and Jankowski [15] measured and compared the strain-rate dependence of the tensile strength on the grain size in crystalline nickel foils, using both tensile and micro-scratch methods. However, it is controversial to correlate the constitutive behaviour measured by tensile tests with the material behaviour measured by nanoindentations, despite the intensive investigations of mechanical properties measured using the instrumented nanoindentation approach [16–20]. In terms of the equivalence of measured material constitutive relationships, it is rare to further quantitatively investigate the constitutive behaviour obtained from tension and indentation methods.

This discrepancy in stress–strain relationships obtained from tension and indentation experiments motivated this paper, in order to find a reliable approach for consistently estimating the constitutive properties from nanoindentation experiments, by calibrating the involved material parameters against the results from tensile experiments. As the aim of this study, the constitutive behaviour measured by nanoindentations is convincingly used for finite element simulations, to examine the mechanical reliability of electronic packaging structures rather than performing tension experiments with the time-consuming preparation of tension samples, which are probably of a size that does not comply with the actual applications. By emphasizing the electronic packaging applications, a representative, lead-free Sn–3.0Ag–0.5Cu (in wt. %; SAC305) solder material is employed to provide a uniform matrix under nanoindentation, in order to rule out the microstructure effect. The solder alloy SAC305 is a typical strain-rate-sensitive, visco-plastic material, and deemed as one of the potential alternatives for consumer electronics [21,22]. Nevertheless, it should be pointed out that the proposed approach herein is generalized for metals and alloys, provided that the prepared material samples have a dense microstructure without significant residual stress.

2. Sample Preparation and Experimental Setup

The SAC305 bulk solder alloy is manufactured by Alpha Assembly Solutions (South Plainfield, NJ, USA) to be free of cast in impurities or included oxides. The SAC305 solder samples for tensile and nanoindentation experiments were prepared in the form of a dog bone and mounted plate, as shown in Figure 1a,b, respectively. The experimental equipment were a Bose ElectroForce 3330 mechanical test machine and a Nano Indenter G200 by Agilent Technologies (Santa Clara, CA, USA). The bulk solder was machined to achieve the desired dog-bone type of specimen shown in Figure 1a, which was designed by referring to American Society for Testing and Materials (ASTM) E8/E8M [23]. The samples for nanoindentations were approximately 10.0 mm × 10.0 mm × 2.0 mm, and were mounted in polyvinyl chloride (PVC) tubes by dental base acrylic resin powder, as shown in the magnified inset of Figure 1b. Despite being from the same material source, thermal treatment was applied in order to align the material property for both types of specimens, by using an air furnace with a temperature stability of ±1.0 °C. This thermal treatment is important for avoiding the effect of micro-defects and residual stress on the mechanical properties of the material sample of interest when correlating the experimental results from tensile and nanoindentation experiments. As found by the authors [24,25], the thermal treatment minimizes the residual stress and stabilizes the mechanical property of the annealed solder. In this study, the applied annealing temperatures were 80 °C, 125 °C, 165 °C, and 210 °C, and the durations were 2 h, 6 h, 12 h, 24 h, and 48 h. It should be noted that the high-temperature annealing temperature at 210 °C is slightly lower than the melting point T_m of 217 °C for SAC305 solder.

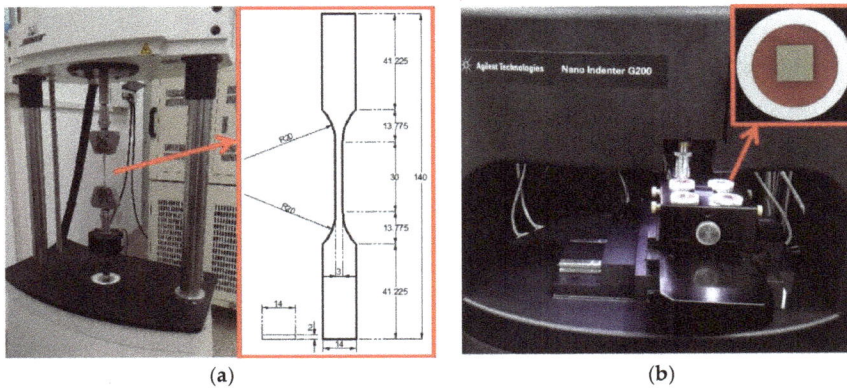

Figure 1. Experimental equipment and material sample. (**a**) Tensile experiment (unit: mm); (**b**) nanoindentation experiment.

The applied strain rate of tensile experiments is 5×10^{-4} s^{-1} under displacement control, in order to obtain the quasi-static behavior of SAC305 solder material. The stress–strain relationships can be directly obtained and readily adopted for finite element simulations of electronic packaging structures. In nanoindentation experiments, using a diamond Berkovich indenter (Agilent Technologies, Santa Clara, CA, USA) in the shape of a three-sided pyramid, the strain rate of 0.05 s^{-1} is applied, with the maximum indentation depth of 2000 nm, to eliminate the influence of surface roughness. It should be noted that if a much lower strain rate is applied, to be compatible with the value of tensile experiments, an indentation will take a few hours or even more. This is extremely time-consuming, and also significantly deteriorates the accuracy of nanoindentation results (such as the determination of the contact area), due to the limitation of thermal drift correction of the nanoindentation instrument.

For each indentation, the applied load–indentation depth response can be divided into three stages, as shown in Figure 2—that is, the loading, holding, and unloading stages. By controlling the indenter speed at various penetration depths, the strain rate of 0.05 s^{-1} is maintained at the temperature of 27 °C, until the maximum depth of 2000 nm. Later, the obtained load–depth curves for repeated experiments are averaged to objectively measure the mechanical properties of SAC305 material. Figure 2 shows that the highlighted area *W* of loading stage can be calculated by integrating the response during the penetration between 0 nm and 2000 nm, and also the contact stiffness *S* can be determined from the initial slope of the applied load–indentation depth response.

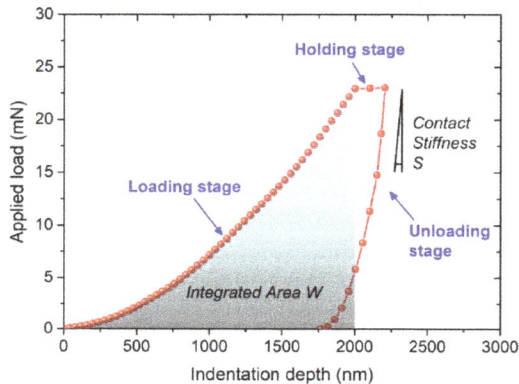

Figure 2. Typical nanoindentation response in the form of applied load–indentation depth.

3. Experimental Results

3.1. Averaged Nanoindentation Response

At least five indentations were performed for each sample among the various annealing treatments. The averaged nanoindentation responses, in the form of applied load–indentation depth curves, are summarized in Figure 3. It was found that annealing treatment is capable of affecting the mechanical behaviour of material under indentation. According to the findings for other metal and alloy materials [26,27], residual stress can be eliminated to a certain extent after the thermal treatment at a high temperature, which will stabilize the microstructure and also the subsequent mechanical properties. Consistent with findings in the literature, the annealing temperature of 210 °C in this study led to a consistent reduction of nanoindentation response (as shown in Figure 3) compared with the unannealed and other annealing temperature conditions, because the applied temperature of 210 °C is closer to the melting point (i.e., 217 °C for SAC305 solder). Additionally, it was also observed that a longer thermal treatment is more effective. Therefore, it is believed that the annealing effect results from thermal accumulation in the form of an equivalent mass diffusion, if the given temperature is sufficiently high. The input energy promotes the alleviation of micro-scale defects to achieve a homogenous eutectic microstructure. Meanwhile, the induced increase of grain size—and thus, a coarser microstructure—will decrease the resistance to dislocation motion, lower yield strength, and working hardening rate [28,29].

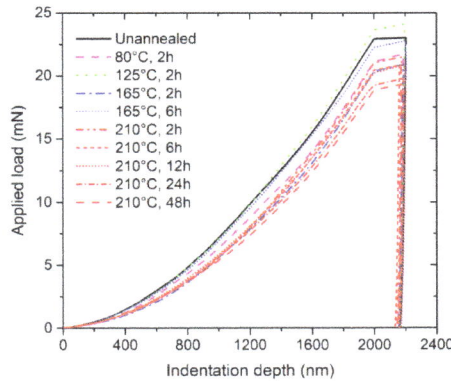

Figure 3. Averaged applied load–indentation depth response to nanoindentation.

3.2. Young's Modulus and Hardness

Based on the continuous stiffness measurement [30], Young's modulus and hardness can be measured as functions of indentation depths, by superimposing a small oscillation to the indentation load controlled by a frequency-specific amplifier to ensure constant amplitude and driving frequency. Based on the harmonic oscillator, the stiffness K_c is provided in Equation (1) by Hay et al. [30]:

$$K_c = 1 / \left[\frac{1}{(F_0/z_0) \cos \phi - (F_0/z_0) \cos \phi|_{free}} - \frac{1}{K_f} \right] \tag{1}$$

where K_f is the elastic stiffness of the frame; ϕ is the phase angle by which the response lags the excitation; and z_0/F_0 is the dynamic compliance, to represent the ratio of the displacement oscillation to the applied excitation. The subscript *free* indicates that the natural frequency of the indenter is in its free-hanging state, so the term $(F_0/z_0) \cos \phi|_{free}$ can be determined as $K - m\omega^2$, where K is the stiffness of the spring supporting the indenter shaft, m is the indenter mass, and $\omega = 2\pi f$ represents the angular frequency of the indenter oscillates. Therefore, the reduced Young's modulus E_r and hardness H can be determined by Equations (2) and (3) [31]:

$$E_r = \frac{\sqrt{\pi}}{2\beta} \frac{K_c}{\sqrt{A_c}} \tag{2}$$

$$H = \frac{P}{A_c} \tag{3}$$

where $A_c = 24.56h_c^2$, and is the projected contact area at the contact depth h_c; the shape constant is $\beta = 1.034$ for a Berkovich indenter; and P is the indentation load. The Young's modulus E can be further calculated by Equation (4).

$$E = \left(1 - \nu^2\right) / \left[\frac{1}{E_r} - \frac{1 - \nu_d^2}{E_d} \right] \tag{4}$$

where ν is 0.42 for the Poisson's ratio of SAC305, and ν_d and E_d are the Poisson's ratio and Young's modulus of diamond, respectively, for the used Berkovich indenter, and are taken as 0.07 and 1140 GPa.

As shown in Figures 4 and 5, these values of Young's modulus and hardness are stabilized after the initial indentation depth of 1000 nm, with some effects due to subtraction, surface stress, and roughness. In the present study, the Young's modulus and hardness can obtained by averaging the corresponding values between 1000 nm and 1100 nm. As summarized in Figure 6, the Young's modulus seems to be more random compared with hardness. The thermal treatment at a higher temperature intends to decrease hardness significantly and consistently, while the treatment duration does not dominate this effect. However, it should be noted that pile-up deformation of indentations may induce the greater elastic modulus calculated from indentation responses, despite some alleviation approaches to recover the linearity of the indentation load and the square of the indentation depth; these alleviation approaches may be done by removing the initial part of the measured load–indentation depth curves, especially for shallow indentation depths.

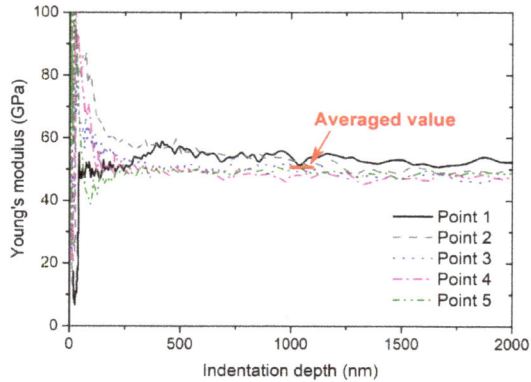

Figure 4. Measured value of Young's modulus as a function of indentation depth.

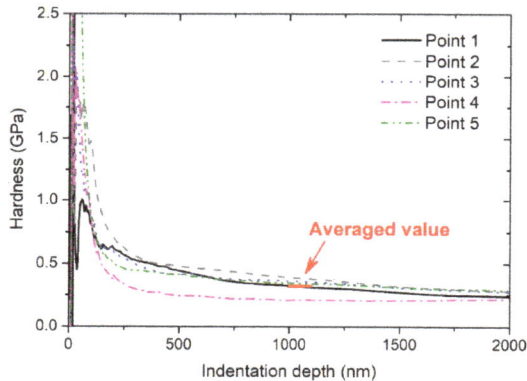

Figure 5. Measured value of hardness as a function of indentation depth.

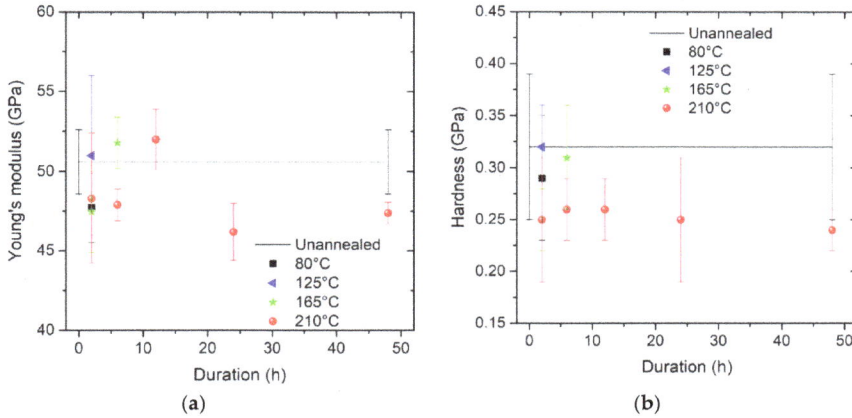

Figure 6. Mechanical properties of the Sn–3.0Ag–0.5Cu (SAC305) solder: (**a**) Young's modulus; (**b**) hardness.

4. Theoretical Analysis

By focusing on the essential relationships between different physical quantities, dimensional analysis has been widely applied in engineering and science areas to identify physical meanings and measurement units, and track these dimensions during formula derivations. In order to reveal the intrinsic mechanism, the dimensionless approach proposed by Ogasawara et al. [32] was adopted herein, due to its fewer parameters with clear physical meanings. The constitutive model in form of a power law $\sigma_R \langle \varepsilon_R \rangle = R(\varepsilon^e + \varepsilon_R)^n$ is parameterized based on dimensionless analysis, in which σ_R is the representative stress; R and n are the hardening rates and exponent, respectively; $\varepsilon^e = \sigma/E$ is the elastic strain; and ε_R is the representative strain, defined as the plastic strain during axisymmetric deformation. The essence of the proposed approach is provided in Equations (5) and (6) for characterizing the dominate information in the loading and unloading parts of the applied load–penetration depth curve, respectively. However, enrichments are made for the parameters σ_R and n that take into account the strain rate effect, as discussed below.

Tensile experiments emphasize the macroscopic-scale deformation behaviour of materials as the average over a great number of microstructural length scales and features, while nanoindentation experiments focus on the local-scale characteristics. In fact, a good agreement can be made between nanoindentation and uniaxial experiments by controlling the indentation strain, using the ratio of loading rate and the applied load proposed by Lucas and Oliver [33]. Atkins and Tabor [34] introduced the concept of representative indentation strain to compare the indentation experiments with uniaxial experiments. They found that the constraint factor is greatly dependent on strain rate, and may lead to a significant discrepancy between the indentation and uniaxial experiments. As compared by Maier et al. [35], the strain rate sensitivity measured by indentation tests is in good agreement with that measured by uniaxial compression tests. Obviously, the strain rates in the tensile and nanoindentation experiments in this study are different, so the strain rate effect is evaluated with further enrichment to unify the constitutive behavior, as shown in Equations (5) and (6) by multiplying the rate factors ψ_σ and ψ_n with the parameters σ_R and n, respectively:

$$\Pi = \frac{W_t}{\delta_{max}^3 \cdot \psi_\sigma \cdot \sigma_R \langle 0.0115 \rangle} = -0.20821\xi^3 + 2.6502\xi^2 - 3.7040\xi + 2.7725 \tag{5}$$

$$\Omega \equiv \frac{S}{2\delta_{max}\overline{E}} = A\xi^3 + B\xi^2 + C\xi + D \tag{6}$$

where $\xi = \ln\left(\overline{E}/\left(\psi_\sigma \cdot \sigma_R \langle 0.0115 \rangle\right)\right)$ with the plane strain modulus $\overline{E} = E/\left(1 - v^2\right)$, Young's modulus Em and Poisson's ratio v; and the representative stress is σ_R, with the representative strain of 0.0115 for a Berkovich indenter. The indentation work done is $W_t = \int_0^{\delta \max} Pd\delta$, determined by area integration from the beginning until the maximum penetration depth of $\delta_{\max}$ in the loading part, and the contact stiffness S is the initial unloading slope of the applied load–penetration depth curve. Both W_t and S have been illustrated in Figure 2. It can be seen in Equation (5) that the maximum penetration depth $\delta_{\max}$ dominates the dimensionless variable Π, which is associated with the loading part. In Equation (6), for the unloading part, the dimensionless variable Ω is a function of $\vartheta = \psi_n \cdot n$, with the hardening exponent of n enriched by the rate factor of ψ_n, in a series of coefficients that are numerically obtained by extensive finite element simulations as follows: $A = -0.04783\vartheta^2 + 0.04667\vartheta - 0.01906$, $B = 0.6455\vartheta^2 - 0.6325\vartheta + 0.2239$, $C = -2.298\vartheta^2 + 2.025\vartheta - 0.4512$, and $D = 2.050\vartheta^2 - 1.502\vartheta + 2.109$.

The value of indentation work done (W_t) and contact stiffness (S) can be determined, as shown in Figure 7, from the applied load–penetration depth curve in Figure 3, as directly recorded from the nanoindentation instrument. Unlike the random distributions for the other thermal treatments, the indentation work done on the samples annealed by the temperature of 210 °C follows a linear relationship with the duration, while the contact stiffness approaches a stable value of about 0.662 with the increasing annealing duration, at the temperature of 210 °C. The unknown variable $\sigma_R \langle 0.0115 \rangle$ and n can be conveniently obtained by finding out the intersection of the two curves from the left and right sides of Equations (5) and (6), respectively. Then, the hardening rate R can be determined by substituting the parameters in a power-law constitutive equation for the representative strain $\varepsilon_R = 0.0115$. It should be noted that a reduction of 30% is made for the value of contact stiffness, as the slope in the initial unloading part is difficult to be quantified, and the automatically recorded values are usually found to be artificially high. This minor assumption does not invalidate the physical basis of the adopted methodology, but ensures the existence of solutions to Equation (6) when solving the hardening exponent n based on Equation (6).

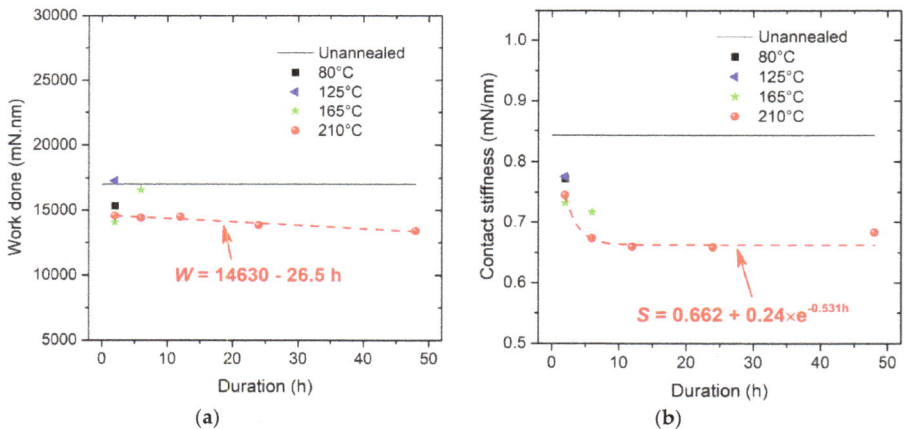

Figure 7. Determined properties from nanoindentation responses: (**a**) contact stiffness during the unloading stage, and (**b**) work done during the loading stage.

The determined values for the representative stress σ_R, the hardening exponent n, and the hardening rate R are provided in Figure 8. Apparently, with increasing duration at the annealing temperature of 210 °C, the representative stress is asymptotically approaching a stabilized value of about 25.23 MPa; the hardening exponent n approximately linearly decreases, and the hardening rate R follows a power-law equation as it decreases. The parameters in Figure 8 are well-described by some fitting formulae, which are therefore inferred to be physically meaningful with regards to dominating the constitutive behaviour.

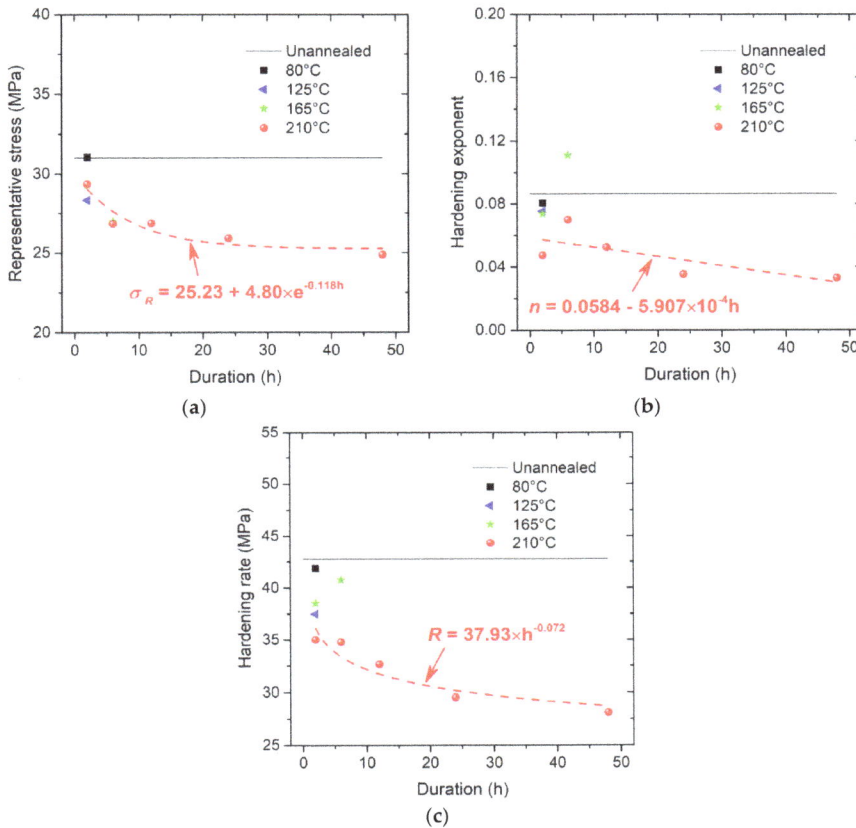

Figure 8. Critical parameters of constitutive model: (**a**) representative stress, (**b**) hardening exponent, and (**c**) hardening rate.

Figures 9 and 10 show that good agreement can be achieved based on comparisons of stress–strain responses measured from tensile experiments and estimated from the applied load–indentation depth responses of nanoindentation experiments. Similar to the published works by Fu et al. [17], there is a certain discrepancy between the predicted and measured curves, especially for the elasto-plastic transition stage. This is very difficult to extract accurately in such a short regime, as explained by Patel and Kalidindi [36]. Nevertheless, in order to best reproduce the stress–strain relationship obtained from the tensile experiment, using the dog-bone type specimens at the strain rate of 5×10^{-4} s^{-1}, the rate factors ψ_σ and ψ_n are determined in Figure 11 to enrich the parameters σ_R and n in the nanoindentation experiments at the strain rate of 0.05 s^{-1}. The general trend of the rate factor ψ_σ is found to be stabilized at the value of 0.52, and the rate factor ψ_n is about 0.10 if the duration is sufficient at the annealing temperature of 210 °C. It is apparent that for both tensile and nanoindentation experiments, the thermal treatments—especially at a sufficiently high temperature for the material sample—are important for stabilizing the mechanical behavior, and align the material property for both types of specimens. Thus, the proposed approach is believed to be reliable for estimating the stress–strain relationships from the nanoindentation responses.

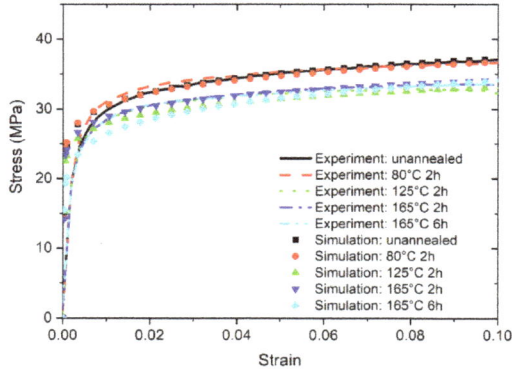

Figure 9. Comparison of stress–strain responses for the annealing temperatures below 210 °C.

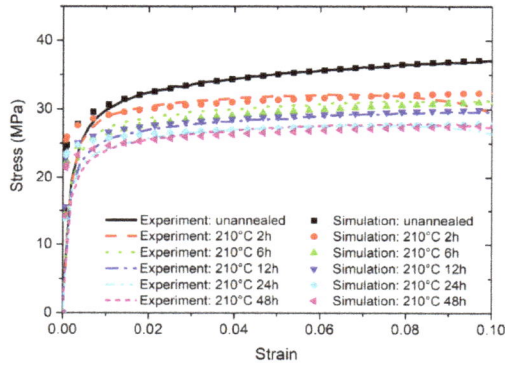

Figure 10. Comparison of stress–strain responses for the annealing temperatures at 210 °C.

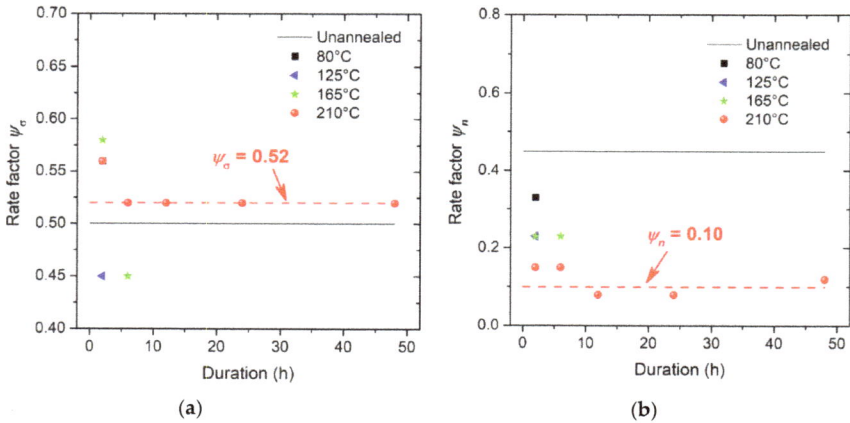

Figure 11. Rate factors for (**a**) representative stress and (**b**) the hardening exponent.

5. Conclusions

In this study, the constitutive behaviour from tensile and nanoindentation experiments was analytically correlated for SAC305 solder samples annealed by various temperatures and durations. Conclusions can be drawn as follows:

- A high annealing temperature close to the melting temperature, with a sufficient duration, benefits the alleviation of residual stress and the stabilization of microstructure, with fewer micro-defects. The constitutive behaviour of SAC305 solder annealed at 210 °C can be used for parameter calibrations.
- Rate factors ψ_σ and ψ_n are proposed and determined to be 0.52 and 0.10, to respectively multiply the representative stress and stress exponent for characterizing the integrated work done and the contact stiffness for the loading and unloading stages of nanoindentation responses.
- The proposed analytical methodology and rate factors can be applicable to other metals and alloys, provided that the material sample of interest is without significant residual stress.

Further studies will be considered to evaluate the generalized rate factors based on the dimensionless approach, so that the stress–strain relationships at a practical range of strain rates can be estimated by performing nanoindentations at a strain rate.

Author Contributions: X.L. conceived and designed the experiments; W.T., S.W., and Y.F. performed the experiments; X.L., X.Z., and C.C. analyzed the data; X.L. wrote the paper.

Acknowledgments: This work was supported by the National Natural Science Foundation of China (No. 51508464), the Natural Science Foundation of Shaanxi Province (No. 2017JM1013), the Astronautics Supporting Technology Foundation of China (No. 2018-HT-XG), and the Fundamental Research Funds for the Central Universities (No. 3102018ZY015).

Conflicts of Interest: The authors declare no conflict of interest.

References

1. Keyes, R.W. The Impact of Moore's Law. *IEEE Solid-State Circuits Soc. Newsl.* **2006**, *11*, 25–27. [CrossRef]
2. Saleh, M.S.; Hu, C.; Panat, R. Three-dimensional microarchitected materials and devices using nanoparticle assembly by pointwise spatial printing. *Sci. Adv.* **2017**, *3*, e1601986. [CrossRef] [PubMed]
3. Frear, D.R. Issues related to the implementation of Pb-free electronic solders in consumer electronics. *J. Mater. Sci. Mater. Electron.* **2007**, *18*, 319–330. [CrossRef]
4. Zhang, L.; Tu, K.N. Structure and properties of lead-free solders bearing micro and nano particles. *Mater. Sci. Eng. R* **2014**, *82*, 1–32. [CrossRef]
5. Xu, S.; Habib, A.H.; Pickel, A.D.; Mchenry, M.E. Magnetic nanoparticle-based solder composites for electronic packaging applications. *Prog. Mater. Sci.* **2015**, *67*, 95–160. [CrossRef]
6. Long, X.; Tang, W.; Feng, Y.; Chang, C.; Keer, L.M.; Yao, Y. Strain rate sensitivity of sintered silver nanoparticles using rate-jump indentation. *Int. J. Mech. Sci.* **2018**, *140*, 60–67. [CrossRef]
7. Zhang, Z.L.; Ni, Y.S.; Zhang, J.M.; Wang, C.; Ren, X.D. Multiscale analysis of size effect of surface pit defect in nanoindentation. *Micromachines* **2018**, *9*, 11. [CrossRef] [PubMed]
8. Bo, W.; Wang, W.; Wang, Y.; Liu, B.; Liu, L. Dynamical modeling and analysis of viscoelastic properties of single cells. *Micromachines* **2017**, *8*, 171.
9. Rengel, M.A.M.; Gomez, F.J.; Rico, A.; Ruiz-Hervias, J.; Rodriguez, J. Obtention of the constitutive equation of hydride blisters in fuel cladding from nanoindentation tests. *J. Nucl. Mater.* **2017**, *487*, 220–228. [CrossRef]
10. Lee, D.-H.; Choi, I.-C.; Yang, G.; Lu, Z.; Kawasaki, M.; Ramamurty, U.; Schwaiger, R.; Jang, J.-I. Activation energy for plastic flow in nanocrystalline CoCrFeMnNi high-entropy alloy: A high temperature nanoindentation study. *Scr. Mater.* **2018**, *156*, 129–133. [CrossRef]
11. Chu, Q.; Zhang, M.; Li, J.; Yan, F.; Yan, C. Investigation of microstructure and fracture toughness of Fe-Zr welded joints. *Mater. Lett.* **2018**, *231*, 134–136. [CrossRef]

12. Hsueh, C.H.; Liao, M.J.; Wang, S.H.; Tsai, Y.T.; Yang, J.R.; Lee, W.S. Size effect and strain induced double twin by nanoindentation in DSS weld metal of vibration-assisted GTAW. *Mater. Chem.Phys.* **2018**, *219*, 40–50. [CrossRef]

13. Schwaiger, R.; Moser, B.; Dao, M.; Chollacoop, N.; Suresh, S. Some critical experiments on the strain-rate sensitivity of nanocrystalline nickel. *Acta Mater.* **2003**, *51*, 5159–5172. [CrossRef]

14. Phani, P.S.; Oliver, W.C. A direct comparison of high temperature nanoindentation creep and uniaxial creep measurements for commercial purity aluminum. *Acta Mater.* **2016**, *111*, 31–38. [CrossRef]

15. Humphrey, R.T.; Jankowski, A.F. Strain-rate sensitivity of strength in macro-to-micro-to-nano crystalline nickel. *Surf. Coat. Technol.* **2011**, *206*, 1845–1849. [CrossRef]

16. Dean, J.; Wheeler, J.M.; Clyne, T.W. Use of quasi-static nanoindentation data to obtain stress–strain characteristics for metallic materials. *Acta Mater.* **2010**, *58*, 3613–3623. [CrossRef]

17. Fu, K.; Chang, L.; Zheng, B.; Tang, Y.; Wang, H. On the determination of representative stress–strain relation of metallic materials using instrumented indentation. *Mater. Des.* **2015**, *65*, 989–994. [CrossRef]

18. Tho, K.K.; Swaddiwudhipong, S.; Liu, Z.S.; Zeng, K. Simulation of instrumented indentation and material characterization. *Mater. Sci. Eng. A* **2005**, *390*, 202–209. [CrossRef]

19. Long, X.; Feng, Y.; Yao, Y. Cooling and annealing effect on indentation response of lead-free solder. *Int. J. Appl. Mech.* **2017**, *9*, 1750057. [CrossRef]

20. Long, X.; Du, C.Y.; Li, Z.; Guo, H.C.; Yao, Y.; Lu, X.Z.; Hu, X.W.; Ye, L.L.; Liu, J. Finite element analysis of constitutive behaviour of sintered silver nanoparticles under nanoindentation. *Int. J. Appl. Mech.* **2018**, *10*, 1–17.

21. Long, X.; He, X.; Yao, Y. An improved unified creep-plasticity model for SnAgCu solder under a wide range of strain rates. *J. Mater. Sci.* **2017**, *52*, 6120–6137. [CrossRef]

22. Long, X.; Tang, W.; Xu, M.; Keer, L.M.; Yao, Y. Electric current-assisted creep behaviour of Sn–3.0Ag–0.5Cu solder. *J. Mater. Sci.* **2018**, *53*, 6219–6229. [CrossRef]

23. Annual Book of ASTM Standards. *Standard Test Methods for Tension Testing of Metallic Materials*; American Association State: West Conshohocken, PA, USA, 2009.

24. Long, X.; Wang, S.; Feng, Y.; Yao, Y.; Keer, L.M. Annealing effect on residual stress of Sn-3.0Ag-0.5Cu solder measured by nanoindentation and constitutive experiments. *Mater. Sci. Eng. A* **2017**, *696*, 90–95. [CrossRef]

25. Long, X.; Wang, S.; He, X.; Yao, Y. Annealing optimization for tin-lead eutectic solder by constitutive experiment and simulation. *J. Mater. Res.* **2017**, *32*, 1–11. [CrossRef]

26. Bai, N.; Chen, X. A new unified constitutive model with short- and long-range back stress for lead-free solders of Sn–3Ag–0.5Cu and Sn–0.7Cu. *Int. J. Plast.* **2009**, *25*, 2181–2203. [CrossRef]

27. Kim, K.S.; Huh, S.H.; Suganuma, K. Effects of cooling speed on microstructure and tensile properties of Sn–Ag–Cu alloys. *Mater. Sci. Eng. A* **2002**, *333*, 106–114. [CrossRef]

28. Ochoa, F.; Williams, J.J.; Chawla, N. Effects of cooling rate on the microstructure and tensile behavior of a Sn-3.5wt.%Ag solder. *J. Electron. Mater.* **2003**, *32*, 1414–1420. [CrossRef]

29. Long, X.; Tang, W.; Wang, S.; He, X.; Yao, Y. Annealing effect to constitutive behavior of Sn–3.0Ag–0.5Cu solder. *J. Mater. Sci. Mater. Electron.* **2018**, *29*, 1–11. [CrossRef]

30. Hay, J.; Agee, P.; Herbert, E. Continuous stiffness measurement during instrumented indentation testing. *Exp. Tech.* **2010**, *34*, 86–94. [CrossRef]

31. Oliver, W.C.; Pharr, G.M. Improved technique for determining hardness and elastic modulus using load and displacement sensing indentation experiments. *J. Mater. Res.* **1992**, *7*, 1564–1583. [CrossRef]

32. Ogasawara, N.; Chiba, N.; Chen, X. Measuring the plastic properties of bulk materials by single indentation test. *Scr. Mater.* **2006**, *54*, 65–70. [CrossRef]

33. Lucas, B.N.; Oliver, W.C. Indentation power-law creep of high-purity indium. *Metall. Mater. Trans. A* **1999**, *30*, 601–610. [CrossRef]

34. Atkins, A.G.; Tabor, D. Plastic indentation in metals with cones. *J. Mech. Phys. Solids* **1965**, *13*, 149–164. [CrossRef]

35. Maier, V.; Durst, K.; Mueller, J.; Backes, B.; Höppel, H.W.; Göken, M. Nanoindentation strain-rate jump tests for determining the local strain-rate sensitivity in nanocrystalline Ni and ultrafine-grained Al. *J. Mater. Res.* **2011**, *26*, 1421–1430. [CrossRef]

36. Patel, D.K.; Kalidindi, S.R. Correlation of spherical nanoindentation stress-strain curves to simple compression stress-strain curves for elastic-plastic isotropic materials using finite element models. *Acta Mater.* **2016**, *112*, 295–302. [CrossRef]

micromachines

MDPI

Article

Nanoindentation of Bi₂Se₃ Thin Films

Hong-Da Lai [1], Sheng-Rui Jian [1,*], Le Thi Cam Tuyen [2], Phuoc Huu Le [3,4,*], Chih-Wei Luo [5] and Jenh-Yih Juang [5]

[1] Department of Materials Science and Engineering, I-Shou University, Kaohsiung 84001, Taiwan; laihongdar95@gmail.com
[2] Department of Materials Science and Engineering, National Chiao Tung University, Hsinchu 30010, Taiwan; ltctuyen89@gmail.com
[3] Theoretical Physics Research Group, Advanced Institute of Materials Science, Ton Duc Thang University, Ho Chi Minh City 700000, Vietnam
[4] Faculty of Applied Sciences, Ton Duc Thang University, Ho Chi Minh City 700000, Vietnam
[5] Department of Electrophysics, National Chiao Tung University, Hsinchu 30010, Taiwan; cwluo@mail.nctu.edu.tw (C.-W.L.); jyjuang@g2.nctu.edu.tw (J.-Y.J.)
* Correspondence: srjian@gmail.com (S.-R.J.); lehuuphuoc@tdt.edu.vn (P.H.L.); Tel.: +886-7-6577711-3130 (S.-R.J.); +028-37755035 (P.H.L.)

Received: 21 August 2018; Accepted: 12 October 2018; Published: 14 October 2018

check for updates

Abstract: The nanomechanical properties and nanoindentation responses of bismuth selenide (Bi₂Se₃) thin films are investigated in this study. The Bi₂Se₃ thin films are deposited on *c*-plane sapphire substrates using pulsed laser deposition. The microstructural properties of Bi₂Se₃ thin films are analyzed by means of X-ray diffraction (XRD). The XRD results indicated that Bi₂Se₃ thin films are exhibited the hexagonal crystal structure with a *c*-axis preferred growth orientation. Nanoindentation results showed the multiple "pop-ins" displayed in the loading segments of the load-displacement curves, suggesting that the deformation mechanisms in the hexagonal-structured Bi₂Se₃ films might have been governed by the nucleation and propagation of dislocations. Further, an energetic estimation of nanoindentation-induced dislocation associated with the observed pop-in effects was made using the classical dislocation theory.

Keywords: Bi₂Se₃ thin films; nanoindentation; hardness; pop-in

1. Introduction

Recently, topological insulators (TIs) have attracted enormous research attention owing to their intriguing fundamental physical properties, such as their conduction mechanisms [1,2], as well as their potential applications in the emergent fields of spintronics [3], optoelectronics [4] and quantum computation [5]. Among various TI materials based on Bi compounds [6,7], bismuth selenide (Bi₂Se₃) is one of the most popular representative candidates in three-dimensional TIs [7,8] suitable for electronic applications, because of its large bulk energy gap of 0.3 eV and a single Dirac cone in the Brillouin zone [1,7]. In addition, Bi₂Se₃ also exhibits excellent thermoelectric properties at roomtemperature [9] and low-temperature regime [10]. For the fundamental study and device application, it is essential to grow Bi₂Se₃ thin films with high-quality and desired mechanical properties [11,12].

Epitaxial Bi₂Se₃ thin films have been successfully prepared by molecular beam epitaxy (MBE) [13–16]. Compared to MBE deposition, pulsed laser deposition (PLD) offers advantages such as a higher instantaneous deposition rate, relatively high reproducibility, and low costs. Thus, PLD has become one of the most widely used deposition techniques for growing thin films containing multi-elements. Both epitaxial and polycrystalline Bi₂Se₃ thin films have been successfully prepared by PLD [9,17–20]. In particular, PLD-grown Bi₂Se₃ thin films on InP (111) substrate presented triangular pyramids with step-and-terrace structures and growth along the [0001] direction [17]. Though lattice misfit over 13%, the Bi_2Se_3 films were epitaxially grown on Al_2O_3 (0001) with in-plane the relationship of (0001) Bi_2Se_3 | | (0001) Al_2O_3 and [$2\bar{1}\bar{1}0$] Bi_2Se_3 | | [$2\bar{1}\bar{1}0$] Al_2O_3 or [$2\bar{1}\bar{1}0$] Bi_2Se_3 | | [$11\bar{2}0$] Al_2O_3 [19]. Meanwhile, the Bi_2Se_3 films prepared by metal organic chemical vapor deposition and thermal evaporation exhibited polycrystalline morphologies and c-axis preferred oriented structures [21,22]. In this study, PLD technique is adopted to grow textured Bi_2Se_3/Al_2O_3 (0001) thin films and study their nanomechanical properties.

The mechanical properties of thin films in nanometer-scale are of great interest since they can be significantly different from their bulk counterparts. Especially, when thin films are used as structural/functional elements of certain nanodevices, robustness to stringent mechanical impacts arising from various fabrication processes is also of pivotal importance. Thus, studies on the correlations between the microstructural and mechanical properties of thin films are indispensable. Nanoindentation has been widely used as a powerful depth-sensing probe for measuring the primary mechanical property parameters, such as hardness and elastic modulus, as well as in revealing the plastic deformation behaviors and mechanisms of various nanoscaled materials [23–26], thin films [27–31] and single-crystal materials [32,33]. Herein, we report the nanomechanical properties of Bi_2Se_3 thin films deposited on *c*-plane sapphire substrates by PLD using nanoindentation with the aid of the continuous contact stiffness (CSM) mode. In addition to obtaining the characteristic nanomechanical properties of Bi_2Se_3 thin films, we also performed detailed analyses on the first pop-in event displayed on the load-displacement curves of nanoindentation to elucidate the underlying plastic deformation mechanisms and the associated dislocation physics [34–37].

2. Materials and Methods

The Bi_2Se_3 thin films investigated in the present study were deposited on Al_2O_3 (0001) substrates by using PLD at a substrate temperature of 300 °C with a helium ambient pressure of 220 mTorr. In particular, in order to obtaining near stoichiometric films at the relatively high substrate temperature of 300 °C, the Se-rich target with a nominal composition of Bi_2Se_8 was used. For the PLD process, ultraviolet (UV) pulses (20-ns duration) from a KrF excimer laser (λ = 248 nm, repetition: 5 Hz) were focused on a polycrystalline Bi_2Se_8 target at a fluence of 6.25 J/cm^2 and a target-to-substrate distance of 40 mm. The deposition time was 20 min, which resulted in an average Bi_2Se_3 film thickness of approximately 360 nm (the growth rate of approximately 0.6 Å/pulse).

The crystalline structure of the obtained Bi_2Se_3 thin films was examined by X-ray diffraction (XRD; Bruker D8, Bruker, Billerica, MA, USA) using theCuKα radiation, λ = 1.54 Å. The surface morphology and film compositions were analyzed by a field emission scanning electron microscopy (FESEM; JEOL JSM-6500, JEOL, Pleasanton, CA, USA) and an Oxford energy-dispersive X-ray spectroscopy (EDS) attached to the SEM instrument, respectively. The analyses were conducted using an accelerating voltage of 15 kV, with the dead time of 22–30% and collection time of 60 s, respectively.

The nanoindentation tests were carried out at a Nanoindenter MTS NanoXP$^{®}$ system (MTS Cooperation, Nano Instruments Innovation Center, Oak Ridge, TN, USA). A three-sided pyramidal Berkovich-type diamond indenter tip with radius of curvature of 50 nm was used for all indentation measurements. The mechanical properties of Bi_2Se_3 thin films were measured by nanoindentation with the continuous contact stiffness (CSM) mode [38]. The indenter was loaded and unloaded three times to ensure that the tip was properly in contact with the material surface, and that any parasitic phenomenon was released from the measurements. Then, the indenter was loaded for the fourth and final time at a strain rate of 0.05 s^{-1}, with a 5 s holding period inserted at the peak load in order to avoid the influence of creep on unloading characteristics, which were used to compute the mechanical properties of Bi_2Se_3 thin films. Finally, the indenter was withdrawn with the same strain rate until 10% of the peak load was reached. At least 20 indents were performed. We also followed the analytic method proposed by Oliver and Pharr [39] to determine the hardness and Young's modulus of Bi_2Se_3 thin films. In order to investigate the cracking phenomenon, cyclic nanoindentation tests were also performed. For the first cycle, the indenter was loaded to some chosen load and then unloaded by 90% of the previous load. It then was reloaded to a larger chosen load and unloaded by 90% for the second cycle. Noticeably, in each cycle, the indenter was hold for 10 s at 10% of its previous maximum load for the thermal drift correction and for assuring unloading completion. The same loading/unloading rate of 10 mN/s was used. The thermal drift was kept below ±0.05 nm/s for all indentations.

3. Results

In Figure 1a, XRD patterns show the dominant (0 0 3*n*) diffraction peaks of Bi_2Se_3 films in addition to a minor Bi_2Se_3 (0 1 5) peak and a Al_2O_3 (0 0 6) peak of the substrate, indicating the film growth along the [0001] direction. This is due to the rhombohedral crystal structure of Bi_2Se_3 (space group $D_{3d}^5(R\overline{3}m)$), in which a hexagonal primitive cell consists of three layers of $-(Se^{(1)}-Bi-Se^{(2)}-Bi-Se^{(1)})$–lamellae (called quintuple layers, QLs) stacking in sequence along the *c*-axis [15]. The interaction between the neighboring QLs is mainly the $Se^{(1)}-Se^{(1)}$ van der Waals bond (Figure 1b). The interlayer $Se^{(1)}-Se^{(1)}$ bonding not only is substantially weaker than the intralayer ionic-covalent bonds within individual QLs but also results in a lowest surface energy on the {001} planes, which leads to observed preferred (001)-oriented crystal growth behavior [9]. As shown the inset of Figure 1a, the full width half maximum (FWHM) of the (0 0 6) peak from the XRD rocking curve was found to be 0.49°, which suggests the presence of certain disorientation between grains (see also Figure 1b). This FWHM was comparable to that of Bi_2Se_3 film grown on Al_2O_3 by PLD [17]. Moreover, the in-plane orientation of the films were examined by XRD Φ-scan on {0 1 5} planes of the Bi_2Se_3 films at a tilt angle (χ) of 57.9°. The films did not show any diffraction peaks, indicating their in-plane polycrystalline characteristics.

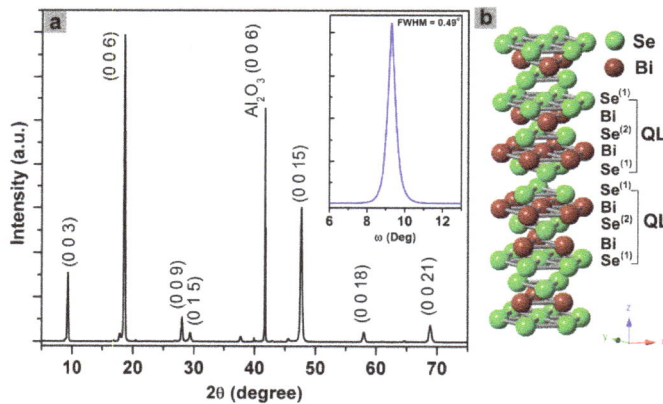

Figure 1. (a) X-ray diffraction (XRD) patterns of a bismuth selenide (Bi_2Se_3) thin film grown on *c*-plane sapphire using pulsed laser deposition (PLD). The inset in (a) shows the XRD rocking curve of (006) peak for the film. (b) Crystal structure of Bi_2Se_3 (QL is quintuple layer).

Intriguingly, the films presented polycrystalline morphology with mutually crossed nanoplatelets (Figure 2a), which are somehow similar to those of Bi_2Te_3 grown by electrodeposition [40]. It has been proposed that the formation of mutually crossed Bi_2Te_3 nanoplatelets can be mainly attributed to the anisotropic bonding nature and growth facet planes with appropriate chemical stoichiometry [40]. This formation mechanism may be also prevailing in the present Bi_2Se_3 films due to the similar anisotropic bonding nature of Bi_2Se_3 and Bi_2Te_3. The film exhibited layered structure and uniform thickness of ~360 nm, as shown by the cross-sectional SEM image in Figure 2a. The upper inset of Figure 2a summarizes the EDS result of the film. Clearly, the film obtained stoichiometric composition of Bi_2Se_3 (i.e., 40.56 at.% Bi and 59.44 at.% Se). The surface roughness can be represented by center line average (R_a), as shown by the AFM image in Figure 2b. The R_a of the film was 8.54 nm.

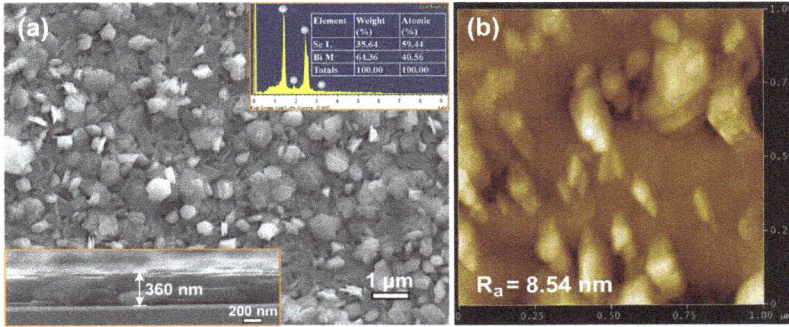

Figure 2. (**a**) A plane-view SEM image of the Bi_2Se_3 thin film deposited on *c*-plane sapphire. Lower inset: a cross-sectional SEM image of the film; upper inset: The energy-dispersive X-ray spectroscopy (EDS) spectra and relative compositions of the film. (**b**) AFM image of the film, R_a is the center line average roughness.

Figure 3a displays the typical load-displacement curve of the present Bi_2Se_3 films obtained by CSM. The corresponding indentation depth-dependent hardness and Young's modulus are shown in Figure 3b,c, respectively. As is evident from Figure 3b, the indentation depth-dependent hardness of Bi_2Se_3 thin film can be roughly divided into two stages. Namely, the hardness, after reaching the maximum in the first 10 nm, precipitously decreases with further increasing indentation depth and eventually reaches a constant value at 2.1 ± 0.1 GPa after the first stage. It is noted that the present results are well within the 30% depth/thickness criterion for nanoindentation test suggested by Li et al. [23,41]. Thus, the effects arising from the substrate or film/substrate interface are excluded. In this respect, the "noisy" depth-dependent hardness, especially in the first stage, might be arisen from the extensive dislocation activities in this stress range. Similar tendency in the depth-dependent Young's modulus is observed (Figure 3c), presumably due to the same mechanism. The Young's modulus of the present Bi_2Se_3 thin film is 58.6 ± 4.1 GPa. It is interesting to note that both the hardness and Young's modulus of present PLD-derived Bi_2Se_3 thin films are much larger than that of single-crystal Bi_2Se_3 reported by Gupta et al. [12], where the respective values of 85.09 MPa and 6.361 GPa were obtained. The reason for the apparent discrepancy is not clear at present. Nevertheless, in addition the apparent differences in microstructure, such as grain boundaries (see Figure 2a), we also note that the load and penetration depths carried out in Reference [12] were both much larger than that employed in the present study. Recently, it has been found in a hybrid double perovskite $(MA)_2AgBiBr_6$ that Young's modulus decreased considerably with increasing indentation depth [42], which partially explains for the larger Young's modulus in this study than that of in Reference [12].

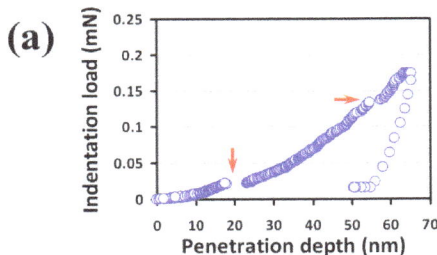

Figure 3. *Cont.*

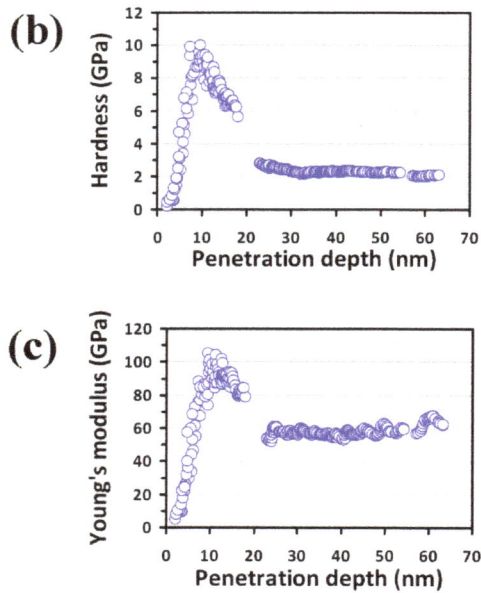

Figure 3. (**a**) A load-displacement curve showing the multiple "pop-ins" during loading part, (**b**) hardness-displacement curve and, (**c**) Young's modulus-displacement curve are obtained from the nanoindentation continuous contact stiffness (CSM) results of Bi_2Se_3 thin film.

From Figure 3a and the cyclic load-displacement curve in the inset of Figure 4, signatures of the multiple pop-ins are clearly observed in the loading part, as indicated by the arrows shown in both figures. It is noted that similar behaviors were also observed in nanoindented Bi_2Se_3 single crystals and was interpreted as being due to heterogeneous nucleation of dislocations beneath the indenter tip [12]. Since the multiple pop-ins is generally closely related to the sudden collective activities of dislocations [43] (such as dislocation generation or movement bursts), we believe that massive dislocation activities are the predominant deformation mechanism in this material, which, in fact, is also consistent with the conjectures of the resultant "noisy" features seen in the depth-dependent curves hardness and Young's modulus described above.

It is also interesting to note that no "pop-out" event is observed in both the unloading curves displayed in Figure 3a and in the inset of Figure 4. Such pop-out behavior is often interpreted as a manifestation of indentation-induced phase transition (for example: nanoindentation-induced phase transformation of single-crystal Si [44]), which is not found in our case. However, as revealed by the SEM image shown in Figure 4, it is evident that significant cracks and pile-ups phenomena along the three corners and edges of the residual indent are also observable. The multiple pop-ins were observed in a large array of materials and were demonstrated to result mainly from massive nucleation and/or propagation of dislocations during loading [45], or micro-cracks initiated around the indentation tip [46]. Hence, it is clear that not only the first pop-in event may reflect the onset of plasticity due to the dislocation activities, but the cracking and pile-up event could also be dominated by the similar mechanism in the present Bi_2Se_3 thin films under nanoindentation. On the other hand, the pressure-induced structural phase transition in Bi_2Se_3 using high pressure Raman and XRD experiments [47] has evidenced that the magnitude of required pressure to induce phase transitions is significantly higher than the apparent room-temperature hardness of hexagonal Bi_2Se_3 thin film measured here. It is worthwhile mentioning that in many hexagonal structured materials, such as, sapphire [48] and GaN thin films [49–52], the primary nanoindentation-induced deformation mechanisms have been consistently identified to be the nucleation and propagation of dislocations. It is, thus, plausible to state that deformation behavior in the present Bi$_2$Se$_3$ thin films is most likely governed by the similar mechanisms.

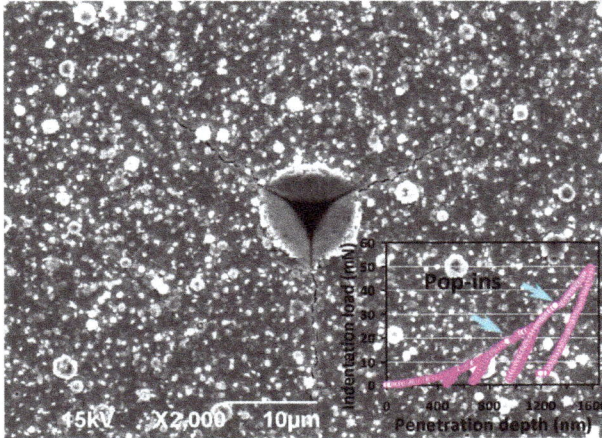

Figure 4. Nanoindented SEM micrograph of Bi$_2$Se$_3$ thin film showing cracks propagate along the corners and pile-up beside the edges of the Berkovich indent. The inset shows the cyclic load-displacement curve at a load of 50 mN. Notice that the multiple "pop-ins" is observable (indicated by the arrows) in loading segments.

Within the scenario of the dislocation nucleation and propagation, the first pop-in event appearing in the loading segment naturally reflects the onset of plasticity for Bi$_2$Se$_3$ thin film, which also provides prominent information about the critical shear stress (τ_{max}) the energy associated with the nucleation of dislocation loops. Following the analytical model proposed by Johnson [53], τ_{max} can be related to the indentation load (P_c), at which a discontinuity in the load-displacement curve takes place, through the following equation [53]:

$$\tau_{max} = \frac{0.31}{\pi}\left[6P_c\left(\frac{E_r}{R}\right)^2\right]^{1/3} \tag{1}$$

Here, R is the radius of indenter tip and E_r is the effective elastic modulus, respectively. The maximum shear stress for Bi$_2$Se$_3$ thin films investigated in the present study is about 0.7 GPa. To the first approximation, the work done by this τ_{max} is mainly associated with the dislocations nucleated within the deformation region underneath the indenter tip. Assuming the nucleation is homogeneous during nanoindentation [34], then, according to the classical dislocation theory [54], the stress at which the first "pop-in" taking place and the energy "dissipated" in it can be regarded, respectively, as the shear stress required to initiate plastic deformation and the energy required for generating a dislocation loop to prevail the deformation. The free energy (U_F) of a circular dislocation loop with radius r can be written as:

$$U_F = \gamma_{dis}2\pi r - \tau b\pi r^2 \tag{2}$$

where γ_{dis} is the energy per unit length of the dislocation loop, b is the magnitude of Burgers vector ($\sim$0.4 nm) [55] and τ is the external shear stress acting on the dislocation loop, respectively. The first term on the right-hand side of Equation (2) describes the energy increased by forming a dislocation loop of radius r in an initially defect-free lattice. The second term is nothing but the strain energy released via work done by the applied stress (τ) to expand the dislocation loop over a displacement of one Burgers vector. The linear energy density (γ_{dis}) for a dislocation is given by [54]:

$$\gamma_{\mathrm{dis}} = \frac{G\,b^2}{8\pi}\left(\frac{2-v_f}{1-v_f}\right)\left[\ln\frac{4r}{r_{\mathrm{core}}}-2\right] \tag{3}$$

where G, v_f and r_{core} are the shear modulus ($\approx$24 GPa), the Poisson's ratio (assumed to be 0.25) of Bi_2Se_3 thin film, and radius of dislocation core, respectively. Substituting Equation (1) and Equation (3) into Equation (2) gives:

$$U_F = \frac{Gb^2r}{4}\left(\frac{2-v_f}{1-v_f}\right)\left(\ln\frac{4r}{r_{\mathrm{core}}}-2\right)-\pi br^2\tau_c \tag{4}$$

Here, τ_c is the resolved shear stress of $\tau_{\max}$ on the active slip systems of the material and is usually taken as half value of $\tau_{\max}$ [56]. Equation (4) clearly indicates that U_F contains terms with first and second power of r. Thus, there must exist a critical radius, r_c, at which U_F of the system reaches a maximum value. When the radius of the dislocation loop exceeds r_c, further expansion lowers U_F, hence is thermodynamically favorable. In contrast, if $r < r_c$, the loop would shrink to reduce the energy. Consequently, when the loading reaches to the "pop-in" point, homogeneous formation of circular dislocation loop becomes possible without thermal energy at $U_F = 0$ [57]. The condition ($U_F = 0$) allows τ_c to be determined from through Equation (2) and Equation (3), yielding $r_c = 2\gamma_{\mathrm{dis}}/(b\tau_{\max})$. Since τ_c has a maximum value as $d\tau_c/dr = 0$, one obtains: $r_c = (e^3 r_{\mathrm{core}})/4$. The values of r_{core} and r_c for the present Bi_2Se_3 thin films were calculated to be 1.08 nm and 5.4 nm, respectively.

By assuming that the nucleation of dislocation loops is entirely responsible for the indentation-induced plastic deformation and no thermal effect is involved, one can further estimate the number of dislocation loops formed during the first "pop-in" event by using the associated work-done (W_p). As depicted in Figure 5, the estimated W_p is ~0.11 $\times$ 10^{-12} Nm, suggesting that ~8 $\times$ 10^3 dislocation loops with critical diameter might have been formed. Although the estimated number is relatively low compared to that of typical polycrystalline thin films (~10^6 cm^{-2}) [58], it is, nevertheless, consistent with the scenario that the "pop-in" is induced by massive homogeneous dislocation nucleation, instead of by the activated collective motion of pre-existing grown-in dislocations [34]. Alternatively, one can take the total dissipation energy as the energy to estimate the number of dislocations with critical radius being generated during entire nanoindentation practice. In that case, as high as ~3 $\times$ 10^5 dislocation loops may be formed during nanoindentation. This number, albeit not entirely realistic, may be considered as the upper limit within the context of dislocation dominant deformation mechanism.

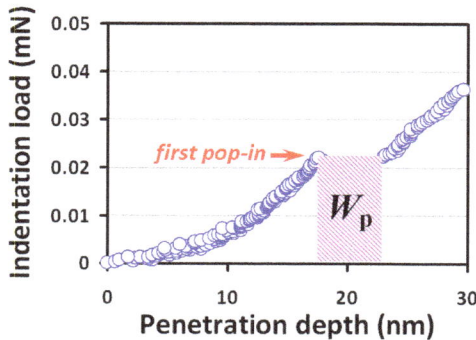

Figure 5. The corresponding first pop-in event from Figure 3a is zoomed in to depict the plastic strain work, W_p, which is approximated as the product of critical loading and the sudden incremental displacement indicated by the shaded area.

4. Conclusions

To sum up, XRD, SEM, AFM and nanoindentation techniques are used to investigate the microstructural and surface morphological features, as well as the nanomechanical properties of Bi_2Se_3 thin films. The results show that the Bi_2Se_3 thin films are polycrystalline with highly (00*l*)-orientation (texture films) and stoichiometric compositions. The hardness and Young's modulus of Bi_2Se_3 thin film are obtained 2.1 ± 0.1 GPa and 58.6 ± 4.1 GPa, respectively. Similar to many hexagonal-structured semiconductors, the primary deformation mechanism for the present Bi_2Se_3 thin film is governed by nucleation and propagation of dislocations or the formation of cracking events. Preliminary energetic estimations indicated that the number of dislocation loops induced by nanoindentation to trigger the plastic deformation accounts for the first pop-in event was in the order of 10^3 with a critical radius ($r_c \approx 5.4$ nm). Although the estimated dislocation density is relatively low compared to that of typical polycrystalline films, it is, nevertheless, in line with the scenario of homogeneous dislocation nucleation-induced first "pop-in" event.

Author Contributions: H.-D.L. contributed to the nanoindentation experiments and analyses. L.T.C.T. and P.H.L. carried out the growth of Bi_2Se_3 thin films and performed XRD and SEM-EDS data. C.-W.L., J.-Y.J. and S.-R.J. contributed to the discussion on materials characterizations. S.-R.J. designed the project of experiments and drafted the manuscript. All authors read and approved the final manuscript.

Acknowledgments: Financial supports from the Ministry of Science and Technology, Taiwan under Contract Nos.: MOST 106-2112-M009-013-MY3, MOST 103-2112-M-009-015-MY3, MOST 107-2112-M-214-001, MOST 106-2112-M-214-001, MOST 105-2112-M-214-001, and Vietnam National Foundation for Science and Technology Development (NAFOSTED) under grant number 103.99–2015.17 are gratefully acknowledged.

Conflicts of Interest: The authors declare no conflicts of interest.

References

1. Zhang, H.; Liu, C.-X.; Qi, X.-L.; Dai, X.; Fang, Z.; Zhang, S.-C. Topological insulators in Bi_2Se_3, Bi_2Te_3 and Sb_2Te_3 with a single Dirac cone on the surface. *Nat. Phys.* **2009**, *5*, 438–442. [CrossRef]

2. Moore, J.E. The birth of topological insulators. *Nature* **2010**, *464*, 194–198. [CrossRef] [PubMed]

3. Yazyev, O.V.; Moore, J.E.; Louie, S.G. Spin Polarization and Transport of Surface States in the Topological Insulators Bi_2Se_3 and Bi_2Te_3 from First Principles. *Phys. Rev. Lett.* **2010**, *105*, 266806. [CrossRef] [PubMed]

4. Min, W.-L.; Betancourt, A.P.; Jiang, P.; Jiang, B. Bioinspired broadband antireflection coatings on GaSb. *Appl. Phys. Lett.* **2008**, *92*, 141109. [CrossRef]

5. Qi, X.-L.; Zhang, S.-C. Topological insulators and superconductors. *Rev. Mod. Phys.* **2011**, *83*, 1057–1110. [CrossRef]

6. Hsieh, D.; Qian, D.; Wray, L.; Xia, Y.; Hor, Y.S.; Cava, R.J.; Hasan, M.Z. A topological Dirac insulator in a quantum spin Hall phase. *Nature* **2008**, *452*, 970–974. [CrossRef] [PubMed]

7. Xia, Y.; Qian, D.; Hsieh, D.; Wray, L.; Pal, A.; Lin, H.; Bansil, A.; Grauer, D.; Hor, Y.S.; Cava, R.J.; et al. Observation of a large-gap topological-insulator class with a single Dirac cone on the surface. *Nat. Phys.* **2009**, *5*, 398–402. [CrossRef]

8. Wiedmann, S.; Jost, A.; Fauqué, B.; van Dijk, J.; Meijer, M.J.; Khouri, T.; Pezzini, S.; Grauer, S.; Schreyeck, S.; Brüne, C.; et al. Anisotropic and strong negative magnetoresistance in the three-dimensional topological insulator Bi_2Se_3. *Phys. Rev. B* **2016**, *94*, 081302(R). [CrossRef]

9. Le, P.H.; Liao, C.-N.; Luo, C.W.; Lin, J.-Y.; Leu, J. Thermoelectric properties of bismuth-selenide films with controlled morphology and texture grown using pulsed laser deposition. *Appl. Surf. Sci.* **2013**, *285*, 657–663. [CrossRef]

10. Hor, Y.S.; Richardella, A.; Roushan, P.; Xia, Y.; Checkelsky, J.G.; Yazdani, A.; Hasan, M.Z.; Ong, N.P.; Cava, R.J. P-type Bi_2Se_3 for topological insulator and low-temperature thermoelectric applications. *Phys. Rev. B* **2009**, *79*, 195208. [CrossRef]

11. Wang, E.; Ding, H.; Fedorov, A.V.; Yao, W.; Li, Z.; Lv, Y.-F.; Zhao, K.; Zhang, L.-G.; Xu, Z.; Schneeloch, J.; et al. Fully gapped topological surface states in Bi_2Se_3 films induced by a d-wave high-temperature superconductor. *Nat. Phys.* **2013**, *9*, 621–625. [CrossRef]

12. Gupta, S.; Vijayan, N.; Krishna, A.; Thukral, K.; Maurya, K.K.; Muthiah, S.; Dhar, A.; Singh, B.; Bhagavannarayana, G. Enhancement of thermoelectric figure of merit in Bi_2Se_3 crystals through a necking process. *J. Appl. Crystallogr.* **2015**, *48*, 533–541. [CrossRef]

13. Chen, J.; Qin, H.J.; Yang, F.; Liu, J.; Guan, T.; Qu, F.M.; Zhang, G.H.; Shi, J.R.; Xie, X.C.; Yang, C.L.; et al. Gate-voltage control of chemical potential and weak anti-localization in bismuth selenide. *Phys. Rev. Lett.* **2010**, *105*, 176602. [CrossRef] [PubMed]

14. Liu, Y.; Weinert, M.; Li, L. Spiral growth without dislocations: Molecular beam epitaxy of the topological insulator Bi_2Se_3 on epitaxial graphene/SiC(0001). *Phys. Rev. Lett.* **2012**, *108*, 115501. [CrossRef] [PubMed]

15. Tarakina, N.V.; Schreyeck, S.; Borzenko, T.; Schumacher, C.; Karczewski, G.; Brunner, K.; Gould, C.; Buhmann, H.; Molenkamp, L.W. Comparative study of the microstructure of Bi_2Se_3 thin films grown on Si(111) and InP(111) substrates. *Cryst. Growth Des.* **2012**, *12*, 1913–1918. [CrossRef]

16. Wang, Z.Y.; Li, H.D.; Guo, X.; Ho, W.K.; Xie, M.H. Growth characteristics of topological insulator Bi_2Se_3 films on different substrates. *J. Cryst. Growth* **2011**, *334*, 96–102. [CrossRef]

17. Onose, Y.; Yoshimi, R.; Tsukazaki, A.; Yuan, H.; Hidaka, T.; Iwasa, Y.; Kawasaki, M.; Tokura, Y. Pulsed Laser Deposition and Ionic Liquid Gate Control of Epitaxial Bi_2Se_3 Thin Films. *Appl. Phys. Express* **2011**, *4*, 83001. [CrossRef]

18. Le, P.H.; Wu, K.H.; Luo, C.W.; Leu, J. Growth and characterization of topological insulator Bi_2Se_3 thin films on $SrTiO_3$ using pulsed laser deposition. *Thin Solid Films* **2013**, *534*, 659–665. [CrossRef]

19. Lee, Y.F.; Punugupati, S.; Wu, F.; Jin, Z.; Narayan, J.; Schwartz, J. Evidence for topological surface states in epitaxial Bi_2Se_3 thin film grown by pulsed laser deposition through magneto-transport measurements. *Curr. Opin. Solid State Mater. Sci.* **2014**, *18*, 279–285. [CrossRef]

20. Orgiani, P.; Bigi, C.; Kumar Das, P.; Fujii, J.; Ciancio, R.; Gobaut, B.; Galdi, A.; Sacco, C.; Maritato, L.; Torelli, P.; et al. Structural and electronic properties of Bi_2Se_3 topological insulator thin films grown by pulsed laser deposition. *Appl. Phys. Lett.* **2017**, *110*, 171601. [CrossRef]

21. Al Bayaz, A.; Giani, A.; Foucaran, A.; Pascal-Delannoy, F.; Boyer, A. Electrical and thermoelectrical properties of Bi_2Se_3 grown by metal organic chemical vapour deposition technique. *Thin Solid Films* **2003**, *441*, 1–5. [CrossRef]

22. Zhang, M. Properties of topological insulator Bi_2Se_3 films prepared by thermal evaporation growth on different substrates. *Appl. Phys. A* **2017**, *123*, 122. [CrossRef]

23. Li, X.; Gao, H.; Murphy, C.J.; Caswell, K.K. Nanoindentation of Silver Nanowires. *Nano Lett.* **2003**, *3*, 1495–1498. [CrossRef]

24. Bao, L.; Xu, Z.-H.; Li, R.; Li, X.D. Catalyst-free synthesis and structural and mechanical characterization of single crystalline $Ca_2B_2O_5.H_2O$ nanobelts and stacking faulted $Ca_2B_2O_5$ nanogrooves. *Nano Lett.* **2010**, *10*, 255–262. [CrossRef] [PubMed]

25. Nagar, R.; Teki, R.; Koratkar, N.; Sathe, V.G.; Kanjilal, D.; Mehta, B.R.; Singh, J.P. Radiation induced modification in nanoscale hardness of ZnO cone structures. *J. Appl. Phys.* **2010**, *108*. [CrossRef]

26. Jian, S.-R.; Sung, T.-H.; Huang, J.C.; Juang, J.-Y. Deformation behaviors of InP pillars under uniaxial compression. *Appl. Phys. Lett.* **2012**, *101*, 151905. [CrossRef]

27. Chen, G.-J.; Jian, S.-R. Effects of Cu doping on the structural and nanomechanical properties of ZnO thin films. *Appl. Phys. A* **2018**, *124*, 575. [CrossRef]

28. Jian, S.-R.; Chen, G.-J.; Lee, J.-W. Effects of annealing temperature on nanomechanical and microstructural properties of Cu-doped In_2O_3 thin films. *Appl. Phys. A* **2017**, *123*, 726. [CrossRef]

29. Jian, S.-R.; Le, P.H.; Luo, C.-W.; Juang, J.-Y. Nanomechanical and wettability properties of Bi_2Te_3 thin films: Effects of post-annealing. *J. Appl. Phys.* **2017**, *121*, 175302. [CrossRef]

30. Le, P.H.; Chiu, S.-P.; Jian, S.-R.; Luo, C.W.; Lin, J.-Y.; Lin, J.-J.; Wu, K.H.; Gospodinov, M. Nanomechanical, structural, and transport properties of Bi_3Se_2Te thin films. *J. Alloys Compd.* **2016**, *679*, 350–357. [CrossRef]

31. Chiu, Y.J.; Shen, C.-Y.; Chang, H.-W.; Jian, S.-R. Characteristics of Iron-Palladium alloy thin films deposited by magnetron sputtering. *Results Phys.* **2018**, *9*, 17–22. [CrossRef]

32. Nawaz, A.; Mao, W.G.; Lu, C.; Shen, Y.G. Mechanical properties, stress distributions and nanoscale deformation mechanisms in single crystal 6H-SiC by nanoindentation. *J. Alloys Compd.* **2017**, *708*, 1046–1053. [CrossRef]

33. Yen, C.-Y.; Jian, S.-R.; Tseng, Y.-C.; Juang, J.-Y. The deformation behavior and fracture toughness of single crystal YSZ(111) by indentation. *J. Alloys Compd.* **2018**, *735*, 2423–2427. [CrossRef]

34. Lorenz, D.; Zeckzer, A.; Hilpert, U.; Grau, P.; Johansen, H.; Leipner, H.S. Pop-in effect as homogeneous nucleation of dislocations during nanoindentation. *Phys. Rev. B* **2003**, *67*, 172101. [CrossRef]

35. Zhu, C.; Lu, Z.P.; Nieh, T.G. Incipient plasticity and dislocation nucleation of FeCoCrNiMn high-entropy alloy. *Acta Mater.* **2013**, *61*, 2993–3001. [CrossRef]

36. Remington, T.P.; Ruestes, C.J.; Bringa, E.M.; Remington, B.A.; Lu, C.H.; Kad, B.; Meyers, M.A. Plastic deformation in nanoindentation of tantalum: A new mechanism for prismatic loop formation. *Acta Mater.* **2014**, *78*, 378–393. [CrossRef]

37. Jian, S.-R.; Ke, W.C.; Juang, J.-Y. Mechanical characteristics of Mg-doped GaN thin films by nanoindentation. *Nanosci. Nanotechnol. Lett.* **2012**, *4*, 598–603. [CrossRef]

38. Li, X.; Bhushan, B. A review of nanoindentation continuous stiffness measurement technique and its applications. *Mater. Charact.* **2002**, *48*, 11–36. [CrossRef]

39. Oliver, W.C.; Pharr, G.M. An improved technique for determining hardness and elastic modulus using load and displacement sensing indentation experiments. *J. Mater. Res.* **1992**, *7*, 1564–1583. [CrossRef]

40. Liao, C.-N.; Shih, H.-D.; Su, P.-W. Electrocrystallization of Mutually Crossed Bismuth Telluride Nanoplatelets. *J. Electrochem. Soc.* **2010**, *157*, D605–D608. [CrossRef]

41. Bhushan, B.; Li, X. Nanomechanical characterisation of solid surfaces and thin films. *Int. Mater. Rev.* **2003**, *48*, 125–164. [CrossRef]

42. Wei, F.; Deng, Z.; Sun, S.; Zhang, F.; Evans, D.M.; Kieslich, G.; Tominaka, S.; Carpenter, M.A.; Zhang, J.; Bristowe, P.D.; Cheetham, A.K. Synthesis and properties of a lead-free hybrid double perovskite: (CH3NH3)2AgBiBr6. *Chem. Mater.* **2017**, *29*, 1089–1094. [CrossRef]

43. Mosca, D.H.; Mattoso, N.; Lepienski, C.M.; Veiga, W.; Mazzaro, I.; Etgens, V.H.; Eddrief, M. Mechanical properties of layered InSe and GaSe single crystals. *J. Appl. Phys.* **2002**, *91*, 140–144. [CrossRef]

44. Jian, S.-R.; Chen, G.-J.; Juang, J.-Y. Nanoindentation-induced phase transformation in (1 1 0)-oriented Si single-crystals. *Curr. Opin. Solid State Mater. Sci.* **2010**, *14*, 69–74. [CrossRef]

45. Jian, S.-R.; Chen, G.-J.; Lin, T.-C. Berkovich nanoindentation on AlN thin films. *Nanoscale Res. Lett.* **2010**, *5*, 935–940. [CrossRef] [PubMed]

46. Bull, S.J. Nanoindentation of coatings. *J. Phys. D. Appl. Phys.* **2005**, *38*, R393–R413. [CrossRef]

47. Yu, Z.; Wang, L.; Hu, Q.; Zhao, J.; Yan, S.; Yang, K.; Sinogeikin, S.; Gu, G.; Mao, H.-K. Structural phase transitions in Bi2Se3 under high pressure. *Sci. Rep.* **2015**, *5*, 1–9. [CrossRef] [PubMed]

48. Nowak, R.; Sekino, T.; Maruno, S.; Niihara, K. Deformation of sapphire induced by a spherical indentation on the (10$\bar{1}$0) plane. *Appl. Phys. Lett.* **1996**, *68*, 1063–1065. [CrossRef]

49. Bradby, J.E.; Kucheyev, S.O.; Williams, J.S.; Wong-Leung, J.; Swain, M.V.; Munroe, P.; Li, G.; Phillips, M.R. Indentation-induced damage in GaN epilayers. *Appl. Phys. Lett.* **2002**, *80*, 383–385. [CrossRef]

50. Jian, S.-R.; Juang, J.-Y. Nanoindentation-induced pop-in effects in GaN thin films. *IEEE Trans. Nanotechnol.* **2013**, *12*, 304–308. [CrossRef]

51. Jian, S.-R. Cathodoluminescence rosettes in c-plane GaN films under Berkovich nanoindentation. *Opt. Mater.* **2013**, *35*, 2707–2709. [CrossRef]

52. Jian, S.-R. Mechanical deformation induced in Si and GaN under Berkovich nanoindentation. *Nanoscale Res. Lett.* **2008**, *3*, 6–13. [CrossRef]

53. Johnson, K.L. *Contact Mechanics*; Cambridge University Press: Cambridge, UK, 1985.

54. Hirth, J.P.; Lothe, J. *Theory of Dislocations*; Wiley: Hoboken, NY, USA, 1981.

55. Zhuang, A.; Li, J.-J.; Wang, Y.-C.; Wen, X.; Lin, Y.; Xiang, B.; Wang, X.; Zeng, J. Screw-dislocation-driven bidirectional spiral growth of Bi2Se3 nanoplates. *Angew. Chemie Int. Ed.* **2014**, *53*, 6425–6429. [CrossRef] [PubMed]

56. Chiu, Y.L.; Ngan, A.H.W. Time-dependent characteristics of incipient plasticity in nanoindentation of a Ni3Al single crystal. *Acta Mater.* **2002**, *50*, 1599–1611. [CrossRef]

57. Leipner, H.S.; Lorenz, D.; Zeckzer, A.; Lei, H.; Grau, P. Nanoindentation pop-in effect in semiconductors. *Physica B* **2001**, *308–310*, 446–449. [CrossRef]

58. Imaizumi, M.; Ito, T.; Yamaguchi, M.; Kaneko, K. Effect of grain size and dislocation density on the performance of thin film polycrystalline silicon solar cells. *J. Appl. Phys.* **1997**, *81*, 7635–7640. [CrossRef]

micromachines

MDPI

Article

Nanoscale-Textured Tantalum Surfaces for Mammalian Cell Alignment

Hassan I. Moussa [1,2]**, Megan Logan** [1,2]**, Kingsley Wong** [1,2]**, Zheng Rao** [1,2]**, Marc G. Aucoin** [1,2] **and Ting Y. Tsui** [1,2,*]

[1] Department of Chemical Engineering, University of Waterloo, Waterloo, ON N2L 3G1, Canada; h2moussa@uwaterloo.ca (H.I.M.); m3logan@uwaterloo.ca (M.L.); kingsley.wong@edu.uwaterloo.ca (K.W.); z2rao@edu.uwaterloo.ca (Z.R.); marc.aucoin@uwaterloo.ca (M.G.A.)

[2] Waterloo Institute of Nanotechnology, University of Waterloo, Waterloo, ON N2L 3G1, Canada

* Correspondence: tttsui@uwaterloo.ca; Tel.: +1-519-888-4567 (ext. 38404)

Received: 12 July 2018; Accepted: 10 September 2018; Published: 13 September 2018

check for updates

Abstract: Tantalum is one of the most important biomaterials used for surgical implant devices. However, little knowledge exists about how nanoscale-textured tantalum surfaces affect cell morphology. Mammalian (Vero) cell morphology on tantalum-coated comb structures was studied using high-resolution scanning electron microscopy and fluorescence microscopy. These structures contained parallel lines and trenches with equal widths in the range of 0.18 to 100 μm. Results showed that as much as 77% of adherent cell nuclei oriented within 10° of the line axes when deposited on comb structures with widths smaller than 10 μm. However, less than 20% of cells exhibited the same alignment performance on blanket tantalum films or structures with line widths larger than 50 μm. Two types of line-width-dependent cell morphology were observed. When line widths were smaller than 0.5 μm, nanometer-scale pseudopodia bridged across trench gaps without contacting the bottom surfaces. In contrast, pseudopodia structures covered the entire trench sidewalls and the trench bottom surfaces of comb structures with line-widths larger than 0.5 μm. Furthermore, results showed that when a single cell simultaneously adhered to multiple surface structures, the portion of the cell contacting each surface reflected the type of morphology observed for cells individually contacting the surfaces.

Keywords: tantalum; mammalian cells; morphology; biomaterials; nanoscale

---

1. Introduction

As a biomaterial [1], tantalum uses include radiopaque bone marker implants and cranioplasty plates [2]. Its alloys have shown promise as orthopedic implant materials due to their osseointegration and bone ingrowth characteristics [3–5]. These metal implants can be used in dense form [6,7] or in porous scaffold structures [4,8–11] for hip and knee arthroplasty [4], spine surgery [4], knee replacement, and avascular necrosis surgery [4,9]. Porous metal scaffolds are used to enhance bone tissue ingrowth and to improve stability performance. The elastic modulus and hardness of 100 nm-thick tantalum thin films are 176.1 ± 3.6 GPa [12] and 12.11 ± 0.46 GPa [12], respectively. Tantalum has a weighted surface energy of ~2.42 J/m^2 [13], which is larger than titanium's weighted surface energy of ~2.0 J/m^2 [13]. Balla et al. [10] showed that human fetal osteoblast cells exhibit better cellular adhesion, growth, and differentiation performance on 73% porous tantalum compared to on titanium control samples. Furthermore, cell densities were six-fold larger on porous tantalum compared to titanium under the same culture conditions. As a result, tantalum thin films are also used to coat porous titanium [14] and carbon scaffold structures [15] to promote implant surface osseointegration and ingrowth characteristics. Although cell responses on bulk specimens are well-established, little knowledge exists about how nanometer-scale textured tantalum surfaces affect cell adhesion and morphology. This information is important as medical implant surfaces may consist of nanometer-scale topographic structures produced during the fabrication processes, for example through mechanical polishing and handling.

The mechanism of cell adhesion and the resulting morphology on different surfaces is complex, often dependent on a wide range of factors such as the protein species adsorbed on the surfaces [16,17], surface structure geometries [17–21], roughness [22–27], and surface energy of the substrata [22,28]. Recently, novel functional biocompatible ferroelectric materials, such as lithium niobate and lithium tantalate, have been used to manipulate cell behavior [29–35]. In particular, the surface charge of these materials is able to enhance osteoblast function, mineral formation [31], and create human neuroblastoma cell patterns [35]. The influences of topographic-based parallel line surface structures on cell adhesion, morphology, and behaviors have been studied by several researchers [36–49]. Some of the literature results for topography-induced morphological changes are summarized in Table 1. Substrate materials used in prior works are limited to polymers, silicon oxide, or silicon. In addition, the range of line width examined in each prior study was often restricted to within two orders of magnitude. The majority of studies thus far have been limited to effects and analysis on a micron scale. There is little information probing effects occurring at or due to sub-micron features. A driving hypothesis of the work presented here is that the range of line widths reported thus far in the literature has limited the ability to gain a full understanding of the effects of surface patterning on cell behavior. However, it is clear from Table 1 that the sensitivity of cell morphology and cell alignment as a result of surface pattern geometries, such as line and trench widths, varies significantly among the cell type and substrate material. No report currently exists regarding the behavior of mammalian cells on nano-textured tantalum surfaces, in part due to the difficulties associated with producing these metal specimens. However, tantalum is increasing in popularity as an implant material. Together with the fact that controlling cell alignment on material surfaces improves the success rate of implants [50–53], there is a need to further understand cell morphology on nano-textured tantalum surfaces.

Table 1. Results of cell alignment performance on various substrate materials and surface pattern designs.

Reference	Cell Type	Substrate	Line Width Range (μm)	Trench Width Range (μm)	Maximum Alignment Line/Trench Width (μm)
[44]	Human corneal epithelial cells	Silicon oxide	0.07–1.9	0.3–2.1	0.85/1.15
[54]	Osteoblast-like cells (MG63)	Silicon	0.09–0.5	0.09–0.5	0.15/0.15
[48]	HeLa cells	Polydimethylsiloxane	2–30	1.5–3.0	2/2
[38]	Human neural stem cells	Polydimethylsiloxane	5–20	5–60	5/5
[37]	Human mesenchymal stem cells	Polystyrene stripes	5–1000	5–1000	20/20
[40]	Adult neural stem cells	Poly-D-lysine with Printed laminin strips	30	30–170	30/30

At the core of this study on cell behavior is how a cell responds to its environment. Cellular organelle-like pseudopodia play an important role in contact guidance, focal adhesion, and motility processes. These cell behaviors are regulated by complex protein-protein interactions and pathways [55,56]. There are wide varieties of pseudopodia and their classifications are commonly based on their morphology, resulting in a sub-classification of filopodia, reticulopodia, and axopodia. These cytoplasmic projections are regulated by different molecular signal transduction pathways.

Hence, the primary objective of this work was to develop an understanding of how complex tantalum-coated nano- and micro-scale comb structures influence mammalian cell morphology and spreading mechanisms. The comb structures included parallel lines and trenches with widths in the range of 0.18 to 100 μm. This study covers more than three orders of magnitude of line/trench widths from nanometer to sub-millimeter scale and is thought to be the largest range by a single investigation to date. Tantalum was chosen for this study in part due to its broad applications in implants [4], mechanical strength [13], corrosive resistance, in vivo bioactivity, and bio-compatibility [57]. Tantalum is even surpassing titanium as a material of choice for certain applications. A secondary aspect enabled by this study was the examination of individual localized responses of cells adhering to multiple patterned tantalum structures having vastly different geometries. In this work, special attention has been given to the behavior of pseudopodia with diameters smaller than 100 nm. Morphology of adherent cells were characterized using high-angle tilted high-resolution field-emission scanning electron microscopy (SEM) and high-resolution fluorescence confocal microscopy techniques. Results showed that cell adhesion and morphology depended not only on line and trench widths but also on the depth pseudopodia penetrated into the trench space. The morphology of an individual cell that simultaneously adhered to different surface pattern structures showed that cells had significantly different localized morphologies and spreading behaviors within the context of a single cell.

2. Materials and Methods

2.1. Tantalum Comb Structures

Tantalum thin film-coated comb structure specimens were fabricated using an advanced integrated circuit back-end-of-line processing method on 200-mm silicon wafers [58–60]. They were supplied by Versum Materials, LLC (Tempe, AZ, USA). The fabrication steps for these silicon-based devices are briefly summarized and illustrated in Figure 1. Parallel line comb structures with equal-width trenches (T) and lines (L) were transferred to the silicon oxide films deposited on the silicon substrate using lithography and dry etching techniques [58–61]. The rectangular-shaped comb structure areas were no smaller than 1.8 mm^2 with widths larger than 1 mm. The tantalum seed layer and copper were deposited on these patterned surfaces, and excess copper was removed using chemical mechanical polish methods [61–63]. The remaining copper was stripped by submerging the specimens in ~9.4 M nitric acid for ~45 min followed by rinsing with deionized water and ethanol. This acid-stripping agent was a diluted solution from 70% nitric acid ACS Plus (Fisherbrand®, Fisher Scientific International Inc., Pittsburgh, PA, USA). The line and trench dimensions fabricated are summarized in Table 2. The trench depths (D) of all patterned comb structures were fixed at ~700 nm.

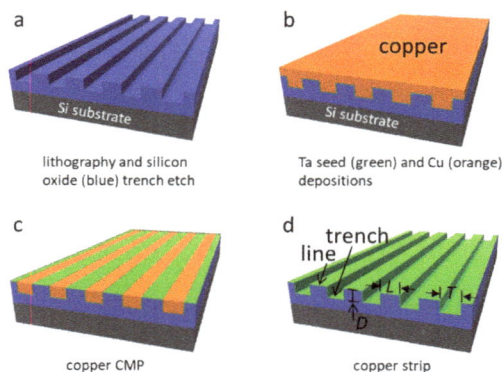

Figure 1. Schematic drawings illustrate the tantalum (green) comb structure fabrication method. (**a**) Patterns were transferred to the silicon oxide films using lithography and plasma etching techniques. (**b**) Tantalum seed and copper films were deposited on the etched patterns. (**c**) Excess copper was removed by using the chemical-mechanical polishing techniques. (**d**) Remaining copper was stripped with nitric acid. The comb structure contains line (L) and trench (T) of equal widths. All trenches had the same depth (D).

Table 2. Data summary of number of cells inspected (n), percent population of cells with $10° > \phi > -10°$ of the line axis, and axis length ratio (L/S). The culture media initial cell concentration used was ~0.5×10^5 cells/mL. Data spreads correspond to one standard deviation.

Structure	Line (L)/Trench (T) Width (µm)	Inspected Comb Structure Area (mm²)	Number of Cells Sampled (n)	Coverage (cell/mm²)	L/S	% of Population Aligned ±10° from Lines
1	0.18	1.8	281	156	2.2 ± 0.7	63.0 ± 1.4
2	0.25	1.8	171	95	2.3 ± 0.8	55.6 ± 4.1
3	0.5	1.8	235	131	2.1 ± 0.6	53.2 ± 7.4
4	1	1.8	197	109	2.7 ± 0.9	77.7 ± 2.0
5	2	1.8	179	99	2.8 ± 1.4	68.7 ± 4.9
6	5	1.8	238	132	2.4 ± 1.2	68.0 ± 6.2
7	10	1.8	159	88	2.3 ± 0.8	71.7 ± 8.6
8	50	1.8	337	187	1.6 ± 0.3	18.7 ± 7.7
9	100	6.6	947	143	1.5 ± 0.4	17.4 ± 0.3
10	blanket Ta	1.8	303	168	1.5 ± 0.4	8.6 ± 5.5

2.2. Cell Culture and Deposition

Detailed cell culturing techniques have been presented elsewhere [18]. Briefly, Vero cells (CCL-81) acquired from the American Type Culture Collection (ATCC, Manassas, VA, USA) were cultured in an equal volume of F12 (Corning, NY, USA) media and Corning®Cellgro™ Dulbecco's Modified Eagle Media (DMEM). The media was supplemented with 4-mM L-glutamine (Sigma-Aldrich, St. Louis, MO, USA) and Gibco™ 10% (v/v) fetal bovine serum (FBS) by Thermo Fisher Scientific (Waltham, MA, USA). Cell culture was performed in 25 mL media under 5% CO_2 atmosphere at 37 °C using tissue-culture-treated 175 cm² flasks (Corning Falcon, Corning, New York, NY, USA). Before inoculation with cells, copper-stripped specimens were sterilized with a 70% ethanol solution for 30 s. This was followed with a Dulbecco's phosphate-buffered saline (D-PBS) rinse. Unless otherwise noted, copper-stripped specimens were then inoculated with ~0.5×10^5–~1.0×10^5 cells/mL and incubated in 6-well tissue culture plates (Nunc, Thermo Scientific, Hvidovre, Denmark) at 37 °C for 0.5 to 24 h.

2.3. Cell Fixation and Staining Processes

All tantalum specimens with adherent cells were rinsed with a D-PBS solution after the prescribed length of incubation and fixed with a solution of 4% methanol-free formaldehyde (Sigma-Aldrich, Oakville, ON, Canada) for 1 h in ambient conditions. The fixed cells were permeabilized in a 0.1% Triton-X 100 (Sigma-Aldrich) solution for 5 min. Specimens were rinsed with PBS and blocked with 2 mL of 1% (w/w) bovine serum albumin (BSA) (Sigma-Aldrich). F-actin microfilament staining was conducted by soaking specimens for 1 h in the deep red CytoPainter F-Actin stain (ab112127 Abcam, Cambridge, MA, USA) solution, which was diluted by a factor of 1000 in 1% BSA. A solution of 0.4 µg/mL of the 4',6-diamidino-2-phenylindole (DAPI, Life Technologies, Waltham, MA, USA) was used to stain the DNA (5 min). All staining processes were performed in the dark to avoid photobleaching and the specimens were rinsed twice with 2 mL D-PBS after each stain application. The final solution contained four drops of Prolong Gold anti-fade reagent (Life Technologies). Specimens were kept refrigerated at 4 °C. A Leica TCS SP5 confocal fluorescence microscope (Wetzlar, Germany) at the University of Guelph, Ontario, Canada, was used to inspect stained samples with wavelengths in the range of 436 to 482 nm (for DAPI) and 650 to 700 nm (for CytoPainter F-Actin).

2.4. Scanning Electron Microscopy

Prior to the SEM inspections, formaldehyde-fixed specimens were dehydrated by soaking them successively in ethanol solutions with increasing concentration: 50%, 75%, 95%, and 100% (v/v). Specimens soaked in the 50% and 75% ethanol were kept in the solution for 10 min each. The final drying processes were completed by two 10-min soaking steps in each of the 95% and 100% solutions. Specimens were dried and then stored in a nitrogen box. Cell cross-sectioning was conducted by using a three-point bend micro-cleaving technique under ambient conditions. Cell inspection and imaging were carried out with a field-emission scanning electron microscope (SEM, Zeiss 1550, Carl Zeiss AG, Oberkochen, Germany). The accelerated voltage was maintained at 7 kV. None of the SEM specimens were coated with gold or other conducting materials.

2.5. Adherent Cell Alignment and Elongation Characterizations

The orientations of adherent cells were characterized by the angles (ϕ) between the long axis of the cell nuclei and the comb structure line axes, as schematically illustrated in Figure 2. The angle of a nucleus' long axis is 90° when it is normal to the line axes (y-axis), whereas the nucleus is aligned parallel to the lines at an angle of 0°. The amount of nuclear elongation was characterized by the ratios between the dimension of the elliptical-shaped nuclei along the long (L) and short (S) axes. Elongated cell nuclei have large L/S values, whereas cells with perfect circular geometry have length ratios of 1. These parameters were manually measured using the built-in functions, Angle and Straight, of Image Processing and Analysis in Java (ImageJ) software (National Institute of Mental Health, Bethesda, MD, USA). To prevent possible influence from the edges of the patterned regions, only measurements recorded from cells that were located further than 50 μm from the perimeter were included in the analyses.

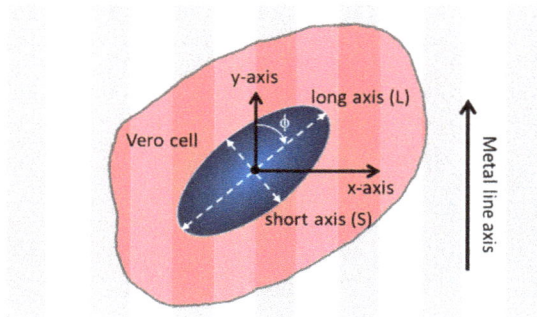

Figure 2. Schematic drawing of a cell on patterned comb structure and their orientation and elongation parameters.

3. Results

3.1. Test Structure Characterizations

Representative 70° tilted SEM micrographs of copper-stripped test structures with varying line widths and spacing are displayed in Figure S1. The entire surface of the specimen, including sidewalls, was coated with a thin (~20 nm) conformal layer of tantalum. The micrographs revealed that all the copper had been removed and the trench side walls were vertically aligned with the substrate surfaces. Both line and trench bottom surfaces were smooth without any observable processing residues. The trenches were approximately 700 nm deep.

3.2. Cell Alignment and Elongation on Patterned Comb Structures

Representative top-down SEM micrographs of adherent cells on the comb structures and blanket tantalum thin film surfaces are displayed in Figure 3. These cells were incubated on these comb structures for 24 h. The comb structures included alternating parallel lines and trenches of the same width. Even though patterns with line widths smaller than 1.0 μm were indiscernible due to the magnifications, the images had lines that were vertically aligned. Micrographs clearly showed that adherent cells on comb structures, having 0.18, 0.25, 0.5, 1.0, 2.0, 5.0, and 10 μm lines (L) and trenches (T) of equal widths, were elongated along the line axes. In contrast, cells on the 50 and 100 µm comb structures maintained arbitrary shapes and did not show any strong orientation preference, and exhibited a similar morphology to that on flat blanket tantalum surfaces. Cell alignment characteristics on the comb structures were also verified with fluorescence confocal microscopy techniques. Micrographs of adherent cells on comb structures with line widths of 0.18, 10, and 50 µm are displayed in Figure 4. The cell nuclei (blue) and F-actin microfilaments (red) were stained with DAPI and phalloidin conjugate, respectively. Results showed that elongated adherent cells and their nuclei were aligned with the line axes on the 0.18 and 10 µm comb structures. In contrast, the majority of cells attached on the 50 µm lines were oriented randomly, similar to those on blanket tantalum thin film surfaces. These observations confirmed that the orientation of the cell nucleus followed the overall adherent cell alignment direction. Additional fluorescence confocal micrographs of cells on 0.25, 0.5, 1.0, and 10 µm comb structures are shown in Figure S2 to demonstrate the reproducibility of the cell morphology. These micrographs show that the cell and nuclear elongation behaviors were consistent with those observed in the SEM micrographs of Figure 3.

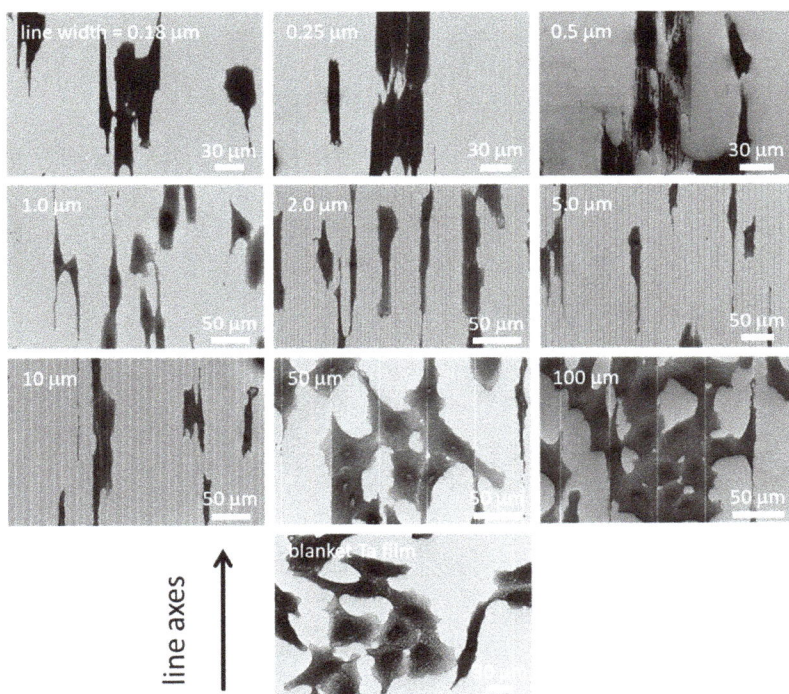

Figure 3. Representative top-down scanning electron microscopy (SEM) micrographs of adherent cells on different comb structures and blanket tantalum (Ta) film. Results show that adherent cells are aligned to the line axes on structures with line widths in the range of 0.18 to 10 µm. In contrast, cells on the 50 µm and 100 µm structures do not align well with the line axes—they are similar to cells randomly distributed on blanket Ta films. All cells were incubated on these specimens for 24 h.

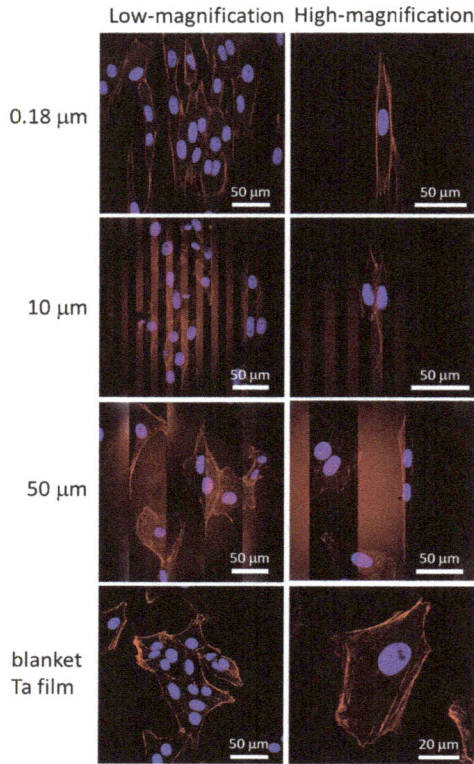

Figure 4. Typical fluorescence confocal micrographs of adherent cells on blanket Ta thin film and comb structures with line widths of 0.18, 10, and 50 μm. Cell nuclei appear blue (4′,6-diamidino-2-phenylindole; DAPI), whereas F-actin microfilaments appear red (fluorescent phalloidin conjugate).

To quantify the cell alignment and elongation behavior, the orientation of the cell's nucleus relative to the line axis (φ) and its dimensions were measured. Figure 5a shows the percentage of the population of cell nuclei that oriented at various angles from the line axes. Specimens with cell nuclei randomly oriented should have an equal distribution in each bin i.e., ~11%. Error bars shown in Figure 5a,b represent one standard deviation from the results of three random groups of cell nuclei. The number of cells and the coverage density (cells/mm^2) of each comb structure are reported in Table 2. Results show that the adherent cells were randomly oriented on the blanket tantalum thin film surfaces with no distinct preferred nuclear orientation. In contrast, cells on the 0.18 to 10 μm comb structures favored alignment parallel to the lines.

To highlight this behavior, the population of cells oriented within ±10° of the line axes is plotted as a function of line width in Figure 5b. Results indicate that there were three possible alignment regimes based on line widths: (i) 0.18 to 0.5 μm, (ii) 1 to 10 μm, and (iii) 50 to 100 μm. In region (i), ~53% to ~63% of the adherent cell population were oriented within the 10° angular range. As line widths increased to between 1 and 10 μm in region (ii), a larger portion of cells, ~68% to ~78%, were aligned with the line axes. Increase in line widths beyond 50 μm, region (iii), led to a sharp decline in cell alignment performance with fewer than 19% of the cell population oriented parallel to the line axes.

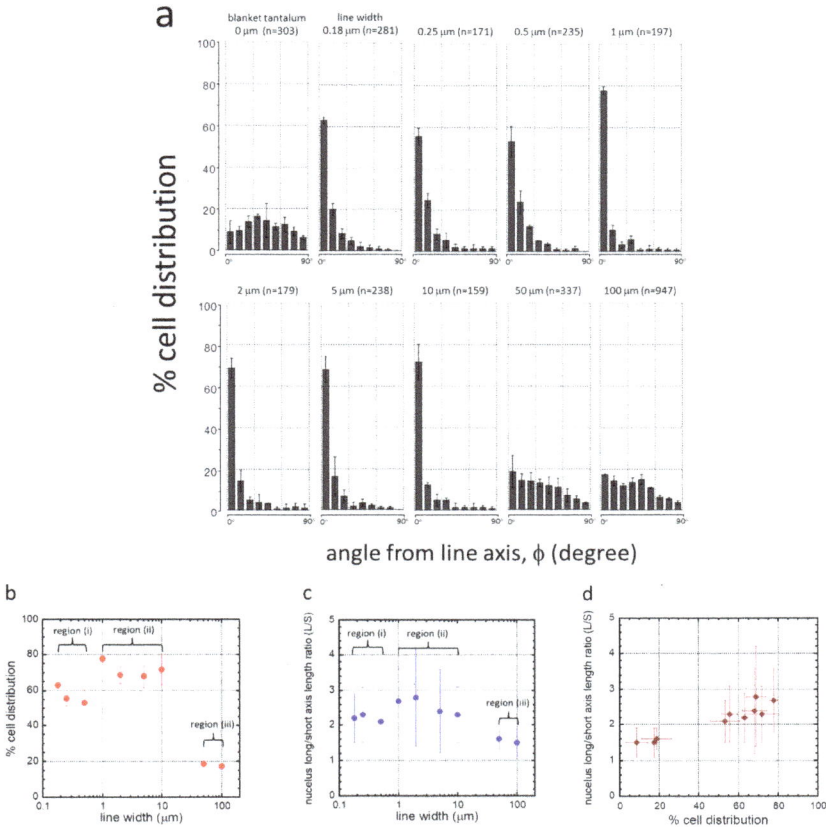

Figure 5. (**a**) Plots of cell orientation (φ) distribution in percentage on 0.18, 0.25, 0.5, 1, 2, 5, 10, 50, and 100 µm-wide line comb structures. As a comparison, this figure includes measurements from blanket tantalum films. The number of cells inspected (n) on each pattern is displayed in an individual chart. Each bar represents a 10° bin of deviations from the line axis in either clockwise or anti-clockwise directions. For example, a cell nucleus deviated from line axis of –22° would fall into the third bin of each plot. These results show that most adherent cells are aligned to the line axes on comb structures in the range of 0.18 to 10 µm. Adherent cells orientations are increasingly randomized on comb structures with line widths of 50 and 100 µm. (**b**) Percent cell distributions that aligned within ±10° of the comb structure line axes. (**c**) Plot of ratio of nucleus long and short axes as a function of comb structure line widths. (**d**) Plot of nucleus axis length ratio (L/S) as a function of percent cell distribution aligned within ± 10° of line axes.

The influence of line width on nuclear elongation was also characterized by comparing the average axis length ratio (L/S) of the cell nuclei on various comb structures, as shown in Figure 5c. Results showed that cell nuclei were significantly elongated when cells were adhered to comb structures with line widths in the range of 0.18 to 10 µm. The largest average length ratio recorded in this range was ~2.8, which occurred on the 2 µm comb structure. In comparison, the length ratio of cell nuclei on the blanket tantalum surface was 1.5 ± 0.4. Nuclei on the 50 and 100 µm comb structures did not exhibit significant elongation with length ratios of ~1.5. The relationship between the axis length ratio and the percentage distribution of cells aligned within 10° of the line axes is shown in Figure 5d. Results showed that as more cell nuclei aligned to the line axes, the average elongation of cell nuclei also increased.

3.3. Nanometer Scale Morphology Analyses

3.3.1. Cells on Individual Comb Structures

Nanometer-scale cell morphology in the three cell alignment regions (i)–(iii) were characterized using high-resolution field-emission SEM techniques. Field-emission SEM was chosen because of the resolution that could be achieved, i.e., smaller than 2 nm. Typical 70° tilted SEM micrographs of cells on 0.18, 0.25, 0.5, 1.0, 2.0, 5.0, 10, and 50 μm-wide line comb structures are shown in Figure 6. These high-angle tilted micrographs captured the three-dimensional morphology of pseudopodia at the periphery of the cells. Micrographs show nanometer-scale pseudopodia spreading in directions parallel and perpendicular to the line axes on the 0.18 and 0.25 μm comb structures. Short pseudopodia filament-like structures that were oriented perpendicular to the line axes are highlighted with red arrows. Some have diameters in the order of ~50 nm and appear to be floating on the patterned structures. To verify this morphology, cells were cross-sectioned by micro-cleaving and inspected using SEM. Typical 70° tilted micrographs of cross-sectioned cells on 0.18 and 0.25 μm comb structures are shown in Figure 7a,b, respectively. Peripheral pseudopodia, highlighted with red arrows, projected across the trench and adhered to adjacent sidewalls. These structures adhered to locations ~80 nm below the top surface and did not contact the trench bottoms—they formed bridges across the trenches. This type of morphology (denoted as Type 1) was also observed away from the periphery of the cell (~2.5 μm) on the 0.5 μm line comb structure (Figure 7c). On the 0.5 μm line comb structure of the periphery, the cells adhered to both sidewalls and trench bottom. This type of cell behavior, denoted as Type 2, was the only morphology observed in comb structures with line widths larger than 1 μm (Figure 6).

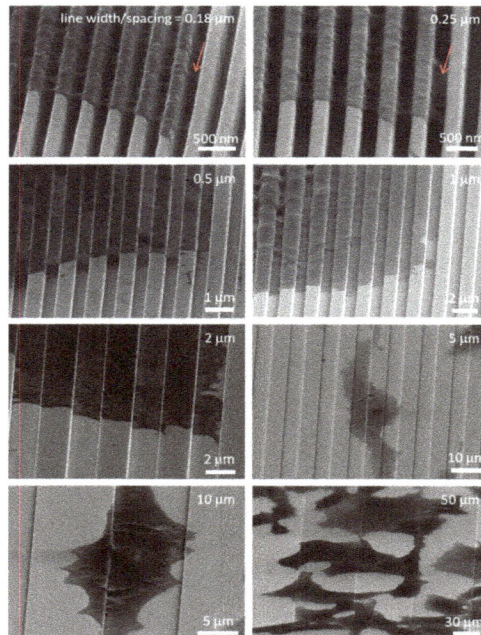

Figure 6. SEM micrographs 70° tilted of cells on comb structures with line widths of 0.18–50 μm. Two distinct types of cell adhesion morphologies are observed: Type 1—Adherent cells on 0.18 μm and 0.25 μm structures only contacted the top portion of lines but did not fill the trench gaps, and Type 2—Cells on comb structures with line widths larger than 1 μm exhibit conformal surface coverages. Both morphology types were observed for adherent cells on the 0.5 μm comb structures. All cells were incubated on the structures for 24 h.

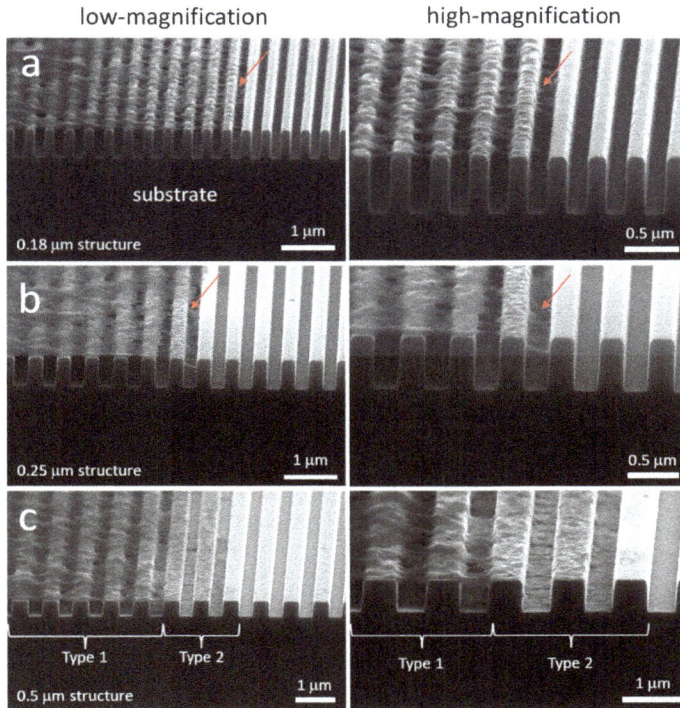

Figure 7. SEM micrographs 70° tilted of cross-sectioned cells on comb structures with line widths of (**a**) 0.18 μm, (**b**) 0.25 μm, and (**c**) 0.5 μm. Two distinct types of cell adhesion morphologies are observed: Type 1—Adherent cells on 0.18 μm and 0.25 μm structures only contacted the top portion of lines but did not fill the trench gaps; and Type 2—Cells exhibit conformal surface coverages. Both morphology types were observed for adherent cells on the 0.5 μm comb structures. All cells were incubated on the structures for 24 h. Cell concentration was ~5 × 10^5 cells/mL.

To determine whether these phenomena occurred during the initial spreading process or only after 24 h of incubation, high-resolution SEM was used to probe the cells on 0.18 and 0.25 μm comb structures after 0.5 and 2 h of incubation. Micrographs in Figure 8 clearly show that the Type 1 pseudopodia morphology occurred as early as 0.5 h post-deposition. The majority of the filament-like structures that were observed bridged across the trenches and adhered to the adjacent sidewalls. Some filament-like structures migrated up the trench sidewalls and covered the line's top surfaces. No cellular material was observed contacting the bottom of the trenches. Additional micrographs of adherent cells on 0.18 and 0.25 μm comb structures after 0.5 h of incubation are shown in Figures S3a and S3b, respectively. The low magnification images of adherent cells reveal that they were not fully spread and had a thick interior region, though the morphology of the peripheral part of the cell was consistent with the images from longer term incubation (Figure 7 and Figure S3c). This morphology does not appear to be the result of the cell contraction process during reshaping as described by others [18].

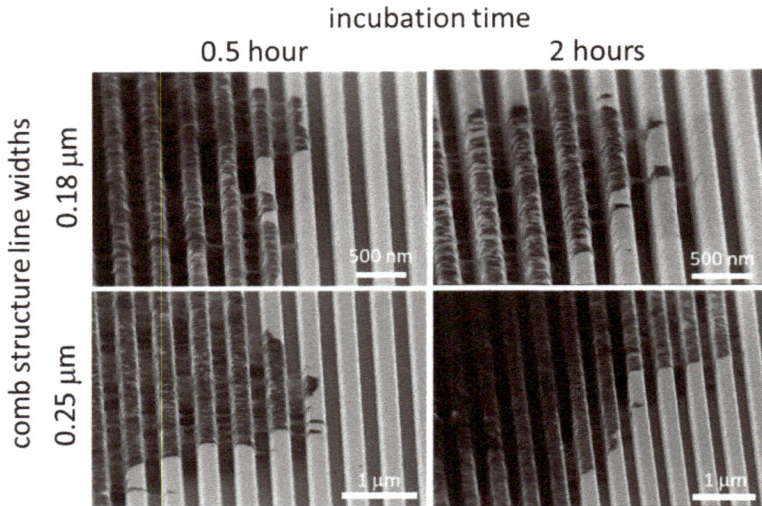

Figure 8. Typical SEM micrographs show Type 1 cell morphology was observed on 0.18 and 0.25 μm comb structures after 0.5 and 2 h of incubation.

The influence of surface topographic parameters, such as trench depth (D), trench width (T), and line width (L), on cell behavior and morphology was also investigated by other researchers [16,37,64–68]. Loseberg et al. [66], Lamers et al. [65], and Ventre et al. [64] showed that fibroblast and osteoblast cell alignment is generally induced by line structures separated by trenches at least 35 nm deep. Lamers and colleagues [65,68], and Toworfe et al. [67] theorized that trenches less than 35 nm deep fill with serum proteins and "smooth" the patterned structure. Line width has also been shown to control cell morphology [16,64–68]. Loseberg et al. [66] and Lamers et al. [65] further reported that line and trench widths larger than 80–100 nm are required to successfully align cells. Our work is consistent with these findings, with cell alignment occurring on patterned surfaces having line widths greater than 180 nm and trench depth of ~700 nm. Trench spacing is also important. Depending on the spacing between lines, cells may conformally coat the trench or bridge across the trench [64]. Epithelial cells deposited on patterns consisting of 330 nm-wide and 150 nm-deep trenches separated by 70 nm lines anchored on the lines and were not able to adhere to the bottom of the trench [16]. In a recent review by Ventre et al. [64], cells are said to "float" on dense patterned structures without contacting the bottom of trenches when line and trench widths are smaller than 100 nm and trench depths are larger than 40 nm. These conditions are also consistent with the observations shown in Figure 6 where cells on comb structures with trench and line widths smaller than 250 nm did not contact the trench bottom.

As trench width increases, cells begin to descend into the trenches [16,45,64] leading to a conformal coating of the surface by the cell. Ventre et al. [64] suggested that a general topographic structural requirement for cells to descend into a trench includes trench widths larger than 1 μm and line widths larger than 100 nm. Zahor et al. [45] showed elongated cells prefer occupying 5 μm-wide trenches rather than the top of lines. Such preferential adhesion is consistent with the results observed in this work (Figures 4 and 6), where the majority of cells deposited on the 10 μm comb structures elongated and descended into the trenches.

3.3.2. A Single Cell Adhered on Multiply Structures Simultaneously

Cell morphology can be influenced by the surface topography of the substrata on which cells can adhere [54,64,69–71]. However, previous observations were based on cells adhered to a uniformly textured surface. It remained unclear how a cell would behave when exposed to a non-uniform textured surface having multiple structures with different geometries. Low magnification top-down and 70° tilted SEM micrographs of cells incubated for 24 h on a smooth blanket tantalum surface and a 0.18 μm comb structure are displayed in Figure 9a,b, respectively. The micrographs indicate that approximately half of the cell and its nucleus adhered to the smooth surface, while the rest of the cell adhered to the comb structure. This is thought to be the first study revealing how mammalian cells can have different morphologies within the same cell when they simultaneously adhere to two different engineered structures. Images showed that cellular materials on the flat surface were spread without preferential orientation. In contrast, the portion of the cell that rested on the patterned line structures elongated and aligned parallel to the line axes. Furthermore, some of the cellular material on the flat surface, adjacent to the comb pattern boundary, appeared to have stretched along the line axes. This may indicate that mechanical stresses were transmitted across the cell; however, the elongation of the cell may have been hindered by the portion of the cell that was anchored on the flat surface. Selected high magnification 70° tilted SEM micrographs of this cell are shown in Figure 9c–f. These micrographs show that the portion of the cell on the comb structure exhibited a Type 1 cell morphology with most cellular materials adhering to the top surfaces of the lines. Nanometer-scale pseudopodia were observed to bridge across the trench gaps as illustrated in Figure 7. These results demonstrate that cells regulated their morphology at a localized level. Additional micrographs of a cell that was incubated on the 0.25 μm comb structures for 0.5 h is shown in Figure S4. The majority of this cell and its nucleus adhered to the flat tantalum surface, while the rest adhered to the comb structures. Type 1 spreading is clearly visible in the comb structure area. However, this cell did not elongate due to the short incubation time.

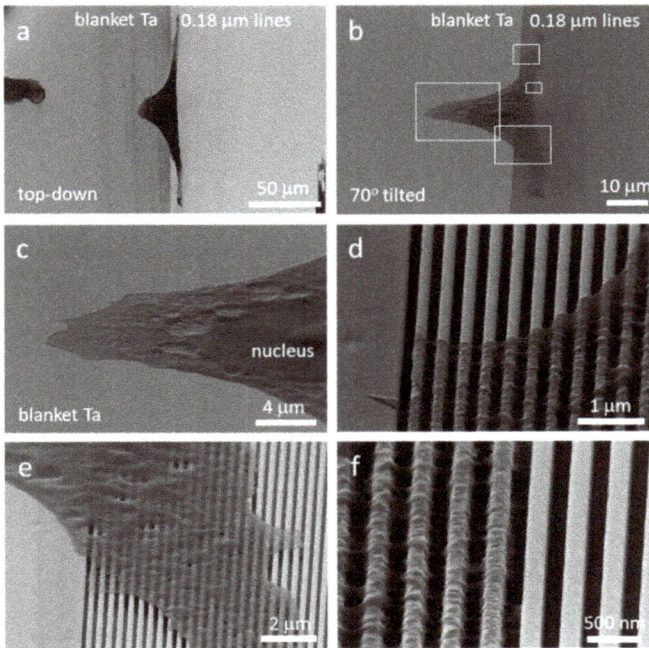

Figure 9. (**a**) Top-down and (**b**) 70° tilted SEM micrographs of an adherent cell partially on the blanket Ta region and on 0.18 μm structures. Portion of cells rested on the blanket tantalum (Ta) region (**c**–**e**) showing regions where cell adhered on blanket Ta and comb structure. (**f**) A high magnification image of cell pattern structure.

3.3.3. Long-Stranded Pseudopodia Structures

In addition to the short pseudopodia filament-like structures shown in Figure 7, micrometer-long single-stranded pseudopodia were also observed. The diameter of these cellular structures was in the order of 50 nm. Representative low- and high-magnification micrographs of these cytoplasmic projections on 0.18 and 0.25 μm comb structures are displayed in Figure 10a,b, respectively. Results showed that these long-stranded pseudopodia bridged across trench gaps on the 0.18 and 0.25 μm comb structures without contacting the trench bottom. This demonstrated that the morphology of long cytoplasmic projections was similar to those short pseudopodia filament-like structures observed at the cell periphery, as shown in Figure 8. Both exhibited a Type 1 structure when cells adhered to comb structures with line widths of 0.18 to 0.25 μm. In addition, Figure 10 shows that the long single-stranded pseudopodia morphology did not change with the distance from the cell periphery.

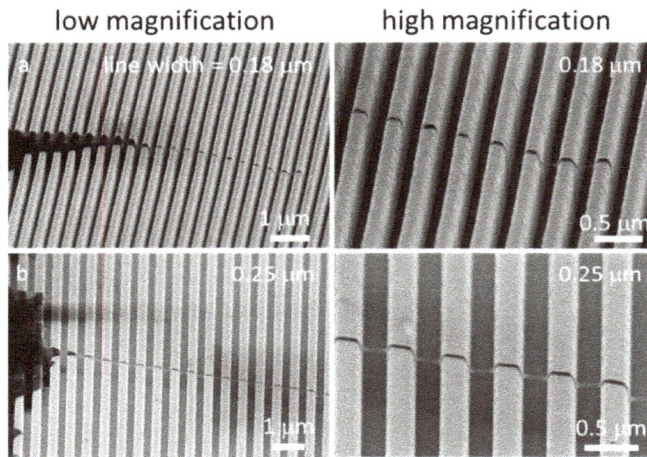

Figure 10. Typical SEM micrographs of adherent cell pseudopodia on (**a**) 0.18 μm and (**b**) 0.25 μm comb structures under low and high magnifications. Type 1 characteristics were observed in all filaments where they wrapped around the top portions of the lines.

3.4. Possible Cell Alignment Mechanisms

One potential contributing factor to the elongated cell morphology shown in Figures 4–7 could have been due to larger mechanical constraints on cell spreading in the direction perpendicular to the lines. When cells spread perpendicular to the line axes, cellular components must overcome the physical constraints by crawling up and down trench sidewalls. These surface topographic features reduce the cell spreading velocity and motility in the perpendicular direction. In contrast, there was no topographic-induced constraint in the direction along the line axes where cells could spread readily. Notably, the amount of mechanical constraint perpendicular to the line axis was pattern-dependent. The number of sidewalls per unit length in the direction perpendicular to the line axes increased with smaller comb structure line widths. Hence, the total distance that cells traveled on dense line comb structure was larger than distances travelled on flat surfaces without any sidewalls, or on comb structures with few sidewalls. Therefore, the amount of cell elongation and alignment was larger on dense line patterns than on flat surfaces. For example, a simple mathematic calculation can show that the total distance a cell spreads perpendicularly across the 1 μm comb structures is ~70% larger than the total distance a cell spreads on the flat surfaces. The slight reduction in cell nuclei alignment performance on 0.18 to 0.5 μm comb structures may seem contradictory to the aforementioned hypothesis; however, this discrepancy may be explained by the different cell-spreading mechanisms (Type 1 vs. Type 2). Figures 7 and 8 show that during the Type 1 spreading process, pseudopodia structures only contacted the top ~80 nm of the trench walls and then bridged across the gaps between adjacent lines. The pseudopodia structures did not cover the entire trench sidewalls and did not contact the trench bottom surfaces. Furthermore, the total distance pseudopodia travelled on the 0.18 μm comb structures was actually ~18% smaller than that travelled on the 1 μm comb structure, but still ~44% greater than that travelled on flat surfaces. Hence, although the cell nuclei on 0.18 μm structures were not as well-aligned as those on the 1 μm comb structure, they performed better than cells on blanket flat surfaces. The cell alignment mechanism proposed here is also consistent with observations by Zhou et al. [41], who reported increases in cell alignment toward the line axes when the cell membrane penetration depths into grooves was larger.

One interesting aspect of the Type 1 cell spreading mechanism is the empty trench space near the cell periphery, which is in the order of hundreds of nanometers wide. These openings create free-standing cell structures that allow a greater surface area to make contact with the surrounding environment, in contrast to cells with the Type 2 spreading mechanism in full contact with the substrata. It is unclear whether these openings produced by the Type 1 morphology could give rise to greater access for contact with nano-particles, or in the context of surgical implants, whether Type I spreading could increase the susceptibility to infection.

In summary, our results demonstrated that patterned tantalum coatings can be used to manipulate cell alignment and morphology. These coatings can potentially be applied to porous scaffold structures as a method to improve matrix material bioactivity or enhance bone regeneration in surgical implants.

4. Conclusions

Adherent mammalian cells (Vero) were elongated on tantalum-coated comb structures with line/trench widths in the range of 0.18 to 10 μm. As much as 77% of the cell nuclei aligned with the line axes. Cell pseudopodia exhibited two types of morphologies that depended on the line and trench widths. First, when widths were smaller than 0.5 μm, nanometer-scale pseudopodia structures bridged across the trenches without contacting the bottom surfaces. Second, cells conformed completely with the surface topology on comb structures having wider line spacing. Results also revealed that individual cells can exhibit multiple morphologies when simultaneously exposed to varying engineered features.

Supplementary Materials: The following are available online at http://www.mdpi.com/2072-666X/9/9/464/s1, Figure S1: Copper stripped comb pattern structures, Figure S2: Fluorescence confocal micrographs of adherent cells on 0.25, 0.5, 1.0, and 10 μm comb structure after 24 hours of incubation, Figure S3: (**a**) 70° tilted SEM micrographs of an adherent cell on 0.18 μm comb structure after 0.5 hour of incubation; (**b**) 70° tilted SEM micrographs of an adherent cell on 0.25 μm comb structure after 0.5 hour of incubation; (**c**) 70° tilted SEM micrographs of an adherent cell on 0.18 μm comb structure after 9 hours of incubation, Figure S4: 70° tilted SEM micrographs of a cell simultaneously adhered on flat surfaces and 0.25 μm comb structure after 0.5 hour of incubation.

Author Contributions: Conceptualization, M.G.A. and T.Y.T.; Data curation, H.I.M., K.W. and Z.R.; Formal analysis, M.G.A. and T.Y.T.; Funding acquisition, T.Y.T.; Investigation, H.I.M., M.L. and T.Y.T.; Methodology, H.I.M., M.L., M.G.A. and T.Y.T.; Supervision, M.G.A. and T.Y.T.; Writing—Original Draft, T.Y.T.; Writing—Review & Editing, M.G.A. and T.Y.T.

Funding: This research was funded by Canadian NSERC Discovery [RGPIN-355552].

Acknowledgments: The authors would like to acknowledge Mark O'Neill of Versum Materials, LLC for support of the chemical-mechanical polished specimens. Ting Y. Tsui thanks Canadian NSERC Discovery [RGPIN-355552] for their support of this work.

Conflicts of Interest: The authors declare no conflict of interest.

References

1.   Kaplan, R.B. Open Cell Tantalum Structures for Cancellous Bone Implants and Cell and Tissue Receptors. U.S. Patent 5,282,861, 1 February 1994.

2. Black, J. Biological Performance of Tantalum. *Clin. Mater.* **1994**, *16*, 173–1994. [CrossRef]
3. Balla, V.K.; Bose, S.; Davies, N.M.; Bandyopadhyay, A. Tantalum—A Bioactive Metal for Implants. *JOM* **2010**, *62*, 61–64. [CrossRef]
4. Levine, B.R.; Sporer, S.; Poggie, R.A.; della Valle, C.J.; Jacobs, J.J. Experimental and clinical performance of porous tantalum in orthopedic surgery. *Biomaterials* **2006**, *27*, 4671–4681. [CrossRef] [PubMed]
5. Miyazakia, T.; Kima, H.-M.; Kokuboa, T.; Ohtsuki, C.; Kato, H.; Nakamura, T. Mechanism of Apatite Formation on Bioactive Titanium Metal TEM-EDX study of mechanism of bonelike apatite formation on bioactive titanium metal in simulated body fluid. *Biomaterials* **2002**, *23*, 3–2002.
6. Varitimidis, S.E.; Dimitroulias, A.P.; Karachalios, T.S.; Dailiana, Z.H.; Malizos, K.N. Outcome after tantalum rod implantation for treatment of femoral head osteonecrosis: 26 hips followed for an average of 3 years. *Acta Orthop.* **2009**, *80*, 20–25. [CrossRef] [PubMed]
7. Ren, B.; Zhai, Z.; Guo, K.; Liu, Y.; Hou, W.; Zhu, Q.; Zhu, J. The application of porous tantalum cylinder to the repair of comminuted bone defects: A study of rabbit firearm injuries. *Int. J. Clin. Exp. Med.* **2015**, *8*, 5055–5064. [PubMed]
8. Tang, Z.; Xie, Y.; Yang, F.; Huang, Y.; Wang, C.; Dai, K.; Zheng, X.; Zhang, X. Porous Tantalum Coatings Prepared by Vacuum Plasma Spraying Enhance BMSCs Osteogenic Differentiation and Bone Regeneration In Vitro and In Vivo. *PLoS ONE* **2013**, *8*, e66263. [CrossRef] [PubMed]
9. Matassi, F.; Botti, A.; Sirleo, L.; Carulli, C.; Innocenti, M. Porous metal for orthopedics implants. *Clin. Cases Miner. Bone Metab.* **2013**, *10*, 111–115. [PubMed]
10. Balla, V.K.; Bodhak, S.; Bose, S.; Bandyopadhyay, A. Porous Tantalum Structures for Bone Implants: Fabrication, Mechanical and In vitro Biological Properties. *Acta Biomater.* **2011**, *6*, 3349–3359. [CrossRef] [PubMed]
11. Bobyn, J.D.; Stackpool, G.J.; Hacking, S.A.; Tanzer, M.; Krygier, J.J. Characteristics of bone ingrowth and interface mechanics of a new porous tantalum biomaterial. *J. Bone Jt. Surg.* **1999**, *81-B*, 907–914. [CrossRef]
12. Guisbiers, G.; Herth, E.; Buchaillot, L.; Pardoen, T. Fracture toughness, hardness, and Young's modulus of tantalum nanocrystalline films. *Appl. Phys. Lett.* **2010**, *97*, 143115. [CrossRef]
13. Tran, R.; Xu, Z.; Radhakrishnan, B.; Winston, D.; Sun, W. Data Descriptor: Surface energies of elemental crystals. *Sci. Data* **2016**, *3*, 1–13. [CrossRef] [PubMed]
14. Wang, Q.; Qiao, Y.; Cheng, M.; Jiang, G.; He, G.; Chen, Y. Tantalum implanted entangled porous titanium promotes surface osseointegration and bone ingrowth. *Nat. Publ. Gr.* **2016**, *6*, 26248. [CrossRef] [PubMed]
15. Wei, X.; Zhao, D.; Wang, B.; Wang, W.; Kang, K.; Xie, H.; Liu, B.; Zhang, X.; Zhang, J.; Yang, Z. Tantalum coating of porous carbon scaffold supplemented with autologous bone marrow stromal stem cells for bone regeneration *in vitro* and *in vivo*. *Exp. Biol. Med.* **2016**, *241*, 592–602. [CrossRef] [PubMed]
16. Teixeira, A.I.; Abrams, G.A.; Bertics, P.J.; Murphy, C.J.; Nealey, P.F. Epithelial contact guidance on well-defined micro- and nanostructured substrates. *J. Cell Sci.* **2003**, *116*, 1892–2003. [CrossRef] [PubMed]
17. Moussa, H.I.; Logan, M.; Siow, G.C.; Phann, D.L.; Rao, Z.; Aucoin, M.G.; Tsui, T.Y. Manipulating mammalian cell morphologies using chemical-mechanical polished integrated circuit chips. *Sci. Technol. Adv. Mater.* **2017**, *18*, 839–856. [CrossRef] [PubMed]
18. Moussa, H.; Logan, M.; Chan, W.; Wong, K.; Rao, Z.; Aucoin, M.; Tsui, T. Pattern-Dependent Mammalian Cell (Vero) Morphology on Tantalum/Silicon Oxide 3D Nanocomposites. *Materials* **2018**, *11*, 1306. [CrossRef] [PubMed]
19. Le Digabel, J.; Richert, A.; Hersen, P.; Ghibaudo, M. Substrate Topography Induces a Crossover from 2D to 3D Behavior in Fibroblast Migration. *Biophys. J.* **2009**, *97*, 357–368.
20. Seo, B.B.; Jahed, Z.; Coggan, J.A.; Chau, Y.Y.; Rogowski, J.L.; Gu, F.X.; Wen, W.; Mofrad, M.R.K.; Tsui, T.Y. Mechanical contact characteristics of pc3 human prostate cancer cells on complex-shaped silicon micropillars. *Materials* **2017**, *10*, 892. [CrossRef] [PubMed]
21. Jahed, Z.; Molladavoodi, S.; Seo, B.B.; Gorbet, M.; Tsui, T.Y.; Mofrad, M.R.K. Cell responses to metallic nanostructure arrays with complex geometries. *Biomaterials* **2014**, *35*, 9363–9371. [CrossRef] [PubMed]
22. Gentleman, M.M.; Gentleman, E. The role of surface free energy in osteoblast—Biomaterial interactions The role of surface free energy in osteoblast—Biomaterial interactions. *Int. Mater. Rev.* **2014**, *59*, 417–429. [CrossRef]

23. Deligianni, D.D.; Katsala, N.; Ladas, S.; Sotiropoulou, D.; Amedee, J.; Missirlis, Y.F. Effect of surface roughness of the titanium alloy Ti-6Al-4V on human bone marrow cell response and on protein adsorption. *Biomaterials* **2001**, *22*, 1251–2001. [CrossRef]

24. Khalili, A.A.; Ahmad, M.R. A Review of Cell Adhesion Studies for Biomedical and Biological Applications. *Int. J. Mol. Sci.* **2015**, *16*, 18149–18184. [CrossRef] [PubMed]

25. Zareidoost, A.; Yousefpour, M.; Ghaseme, B.; Amanzadeh, A. The relationship of surface roughness and cell response of chemical surface modification of titanium. *J. Mater. Sci. Mater. Med.* **2012**, *23*, 1479–1488. [CrossRef] [PubMed]

26. Dolatshahi-Pirouz, A.; Jensen, T.; Kraft, D.C.; Foss, K.M.; Kingshott, P.; Hansen, J.L.; Larsen, A.N.; Chevallier, J.; Besenbacher, F. Fibronectin Adsorption, Cell Adhesion, and Proliferation on Nanostructured Tantalum Surfaces. *ACS Nano* **2010**, *4*, 2874–2882. [CrossRef] [PubMed]

27. Dolatshahi-Pirouz, A.; Pennisi, C.P.; Skeldal, S.; Foss, M.; Chevallier, J.; Zachar, V.; Andreasen, P.; Yoshida, K.; Besenbacher, F. The influence of glancing angle deposited nano-rough platinum surfaces on the adsorption of fibrinogen and the proliferation of primary human fibroblasts. *Nanotechnology* **2009**, *20*, 095101. [CrossRef] [PubMed]

28. Hallab, N.J.; Bundy, K.J.; Connor, K.O.; Moses, R.L.; Jacobs, J.J. Evaluation of Metallic and Polymeric Biomaterial Surface Energy and Surface Roughness Characteristics for Directed Cell Adhesion. *Tissue Eng.* **2001**, *7*, 55–71. [CrossRef] [PubMed]

29. Marchesano, V.; Gennari, O.; Mecozzi, L.; Grilli, S.; Ferraro, P. Effects of Lithium Niobate Polarization on Cell Adhesion and Morphology. *ACS Appl. Mater. Interfaces* **2015**, *7*, 18113–18119. [CrossRef] [PubMed]

30. Christophi, C.; Cavalcanti-Adam, E.A.; Hanke, M.; Kitamura, K.; Gruverman, A.; Grunze, M.; Dowben, P.A.; Rosenhahn, A. Adherent cells avoid polarization gradients on periodically poled LiTaO3ferroelectrics. *Biointerphases* **2013**, *8*, 1–9. [CrossRef] [PubMed]

31. Carville, N.C.; Collins, L.; Manzo, M.; Gallo, K.; Lukasz, B.I.; McKayed, K.K.; Simpson, J.C.; Rodriguez, B.J. Biocompatibility of ferroelectric lithium niobate and the influence of polarization charge on osteoblast proliferation and function. *J. Biomed. Mater. Res. Part A* **2015**, *103*, 2540–2548. [CrossRef] [PubMed]

32. Vilarinho, P.M.; Barroca, N.; Zlotnik, S.; Félix, P.; Fernandes, M.H. Are lithium niobate (LiNbO$_3$) and lithium tantalate (LiTaO$_3$) ferroelectrics bioactive? *Mater. Sci. Eng. C* **2014**, *39*, 395–402. [CrossRef] [PubMed]

33. Mandracchia, B.; Gennari, O.; Marchesano, V.; Paturzo, M.; Ferraro, P. Label free imaging of cell-substrate contacts by holographic total internal reflection microscopy. *J. Biophotonics* **2017**, *10*, 1163–1170. [CrossRef] [PubMed]

34. Mandracchia, B.; Gennari, O.; Bramanti, A.; Grilli, S.; Ferraro, P. Label-free quantification of the effects of lithium niobate polarization on cell adhesion via holographic microscopy. *J. Biophotonics* **2018**, *11*, 1–6. [CrossRef] [PubMed]

35. Rega, R.; Gennari, O.; Mecozzi, L.; Grilli, S.; Pagliarulo, V.; Ferraro, P. Bipolar Patterning of Polymer Membranes by Pyroelectrification. *Adv. Mater.* **2016**, *28*, 454–459. [CrossRef] [PubMed]

36. English, A.; Azeem, A.; Spanoudes, K.; Jones, E.; Tripathi, B.; Basu, N.; Mcnamara, K.; Tofail, S.A.M.; Rooney, N.; Riley, G.; et al. Acta Biomaterialia Substrate topography: A valuable in vitro tool, but a clinical red herring for in vivo tenogenesis. *Acta Biomater.* **2015**, *27*, 12–2015. [CrossRef] [PubMed]

37. Nakamoto, T.; Wang, X.; Kawazoe, N.; Chen, G. Biointerfaces Influence of micropattern width on differentiation of human mesenchymal stem cells to vascular smooth muscle cells. *Colloids Surf. B Biointerfaces* **2014**, *122*, 323–2014. [CrossRef] [PubMed]

38. Béduer, A.; Vieu, C.; Arnauduc, F.; Sol, J.; Loubinoux, I.; Vaysse, L. Biomaterials Engineering of adult human neural stem cells differentiation through surface micropatterning. *Biomaterials* **2012**, *33*, 504–514. [CrossRef] [PubMed]

39. Ferrari, A.; Cecchini, M.; Serresi, M.; Faraci, P.; Pisignano, D.; Beltram, F. Biomaterials Neuronal polarity selection by topography-induced focal adhesion control. *Biomaterials* **2010**, *31*, 4682–4694. [CrossRef] [PubMed]

40. Joo, S.; Kim, J.Y.; Lee, E.; Hong, N.; Sun, W.; Nam, Y. Effects of ECM protein micropatterns on the migration and differentiation of adult neural stem cells. *Sci. Rep.* **2015**, *5*, 13043. [CrossRef] [PubMed]

41. Kim, D.; Provenzano, P.P.; Smith, C.L.; Levchenko, A. Matrix nanotopography as a regulator of cell function. *J. Cell Biol.* **2012**, *197*, 351–360. [CrossRef] [PubMed]

42. Yim, E.K.F.; Reano, R.M.; Pang, S.W.; Yee, A.F.; Chen, C.S.; Leong, K.W. Nanopattern-induced changes in morphology and motility of smooth muscle cells. *Biomaterials* **2005**, *26*, 5405–5413. [CrossRef] [PubMed]

43. Kim, D.; Lipke, E.A.; Kim, P.; Cheong, R.; Thompson, S.; Delannoy, M. Nanoscale cues regulate the structure and function of macroscopic cardiac tissue constructs. *Proc. Natl. Acad. Sci. USA* **2009**, *107*, 565–570. [CrossRef] [PubMed]

44. Teixeira, A.I.; McKie, G.A.; Foley, J.D.; Bertics, P.J.; Nealey, P.F.; Murphy, C.J. The effect of environmental factors on the response of human corneal epithelial cells to nanoscale substrate topography. *Biomaterials* **2006**, *27*, 3945–3954. [CrossRef] [PubMed]

45. Zahor, D.; Radko, A.; Vago, R.; Gheber, L.A. Organization of mesenchymal stem cells is controlled by micropatterned silicon substrates. *Mater. Sci. Eng. C* **2007**, *27*, 121–2007. [CrossRef]

46. Kaiser, J.; Reinmann, A.; Bruinink, A. The effect of topographic characteristics on cell migration velocity. *Biomaterials* **2006**, *27*, 5230–5241. [CrossRef] [PubMed]

47. Fujita, S.; Ohshima, M.; Iwata, H. Time-lapse observation of cell alignment on nanogrooved patterns. *J. R. Soc. Interface* **2009**, *6*, S269–S277. [CrossRef] [PubMed]

48. Zhou, X.; Shi, J.; Hu, J.; Chen, Y. Cells cultured on microgrooves with or without surface coating: Correlation between cell alignment, spreading and local membrane deformation. *Mater. Sci. Eng. C* **2013**, *33*, 855–863. [CrossRef] [PubMed]

49. Tang, Q.Y.; Tong, W.Y.; Shi, J.; Shi, P. Influence of engineered surface on cell directionality and motility. *Biofabrication* **2014**, *6*, 15011. [CrossRef] [PubMed]

50. Jia, M.Z.; Tsuru, K.; Hayakawa, S.; Osaka, A. Modification of Ti implant surface for cell proliferation and cell alignment. *J. Biomed. Mater. Res. Part A* **2008**, *84*, 988–993.

51. Chehroudi, B.; Ratkay, J.; Brunette, D.M. The role of implant surface geometry on mineralization in vivo and in vitro: A transmission and scanning electron microscopic study. *Cells Mater.* **1992**, *2*, 89–104.

52. Owen, G.R.; Jackson, J.; Chehroudi, B.; Burt, H.; Brunette, D.M. A PLGA membrane controlling cell behaviour for promoting tissue regeneration. *Biomaterials* **2005**, *26*, 7447–7456. [CrossRef] [PubMed]

53. Barr, S.; Hill, E.; Bayat, A. Current implant surface technology: An examination of their nanostructure and their influence on fibroblast alignment and biocompatibility. *Eplasty* **2009**, *9*, e22. [PubMed]

54. Yang, J.-Y.; Ting, Y.-C.; Lai, J.-Y.; Liu, H.-L.; Fang, H.-W.; Tsai, W.-B. Quantitative analysis of osteoblast-like cells (MG63) morphology on nanogrooved substrata with various groove and ridge dimensions. *J. Biomed. Mater. Res. A* **2009**, *90*, 629–640. [CrossRef] [PubMed]

55. Nobes, C.D.; Hall, A. Rho, Rac, and Cdc42 GTPases regulate the assembly of multimolecular focal complexes associated with actin stress fibers, lamellipodia, and filopodia. *Cell* **1995**, *81*, 53–62. [CrossRef]

56. Ridley, A.J.; Hall, A. The small GTP-binding protein rho regulates the assembly of focal adhesions and stress fibres in response to growth factors. *Cell* **1992**, *70*, 399–1992.

57. Huo, W.T.; Zhao, L.Z.; Yu, S.; Yu, Z.T.; Zhang, P.X.; Zhang, Y.S. Significantly enhanced osteoblast response to nano-grained pure tantalum. *Sci. Rep.* **2017**, *7*, 1–13. [CrossRef] [PubMed]

58. Doering, R.; Nishi, Y. *Handbook of Semiconductor Manufacturing Technology*, 2nd ed.; CRC Press, Taylor & Francis Group: New York, NY, USA, 2007.

59. Chen, W.-K. *The VLSI Handbook*, 2nd ed.; CRC Press, Taylor & Francis Group: New York, NY, USA, 2007.

60. Li, Y. *Microelectronic Applications of Chemical Mechanical Planarization*; John Wiley & Sons Inc.: Hoboken, NJ, USA, 2007.

61. Van Zant, P. *Microchip Fabrication: A Practical Guide to Semiconductor Processing*, 6th ed.; McGraw Hilll Education: New York, NY, USA, 2014.

62. Shi, X.; Murella, K.; Schlueter, J.A.; Choo, J.O. Chemical Mechanical Polishing Slurry Compositions and Method Using the Same for Copper and Through-Silicon via Applications. U.S. Patent 8,974,692 B2, 10 March 2015.

63. Shi, X.; Palmer, B.J.; Sawayda, R.A.; Coder, F.A.; Perez, V. Method and Composition for Chemical Mechanical Planarization of a Metal. U.S. Patent 8,414,789 B2, 23 September 2013.

64. Ventre, M.; Causa, F.; Netti, P.A. Determinants of cell-material crosstalk at the interface: Towards engineering of cell instructive materials. *J. R. Soc. Interface* **2012**, *9*, 2017–2032. [CrossRef] [PubMed]

65. Lamers, E.; van Horssen, R.; te Riet, J.; van Delft, F.C.; Luttge, R.; Walboomers, X.F.; Jansen, J.A. The influence of nanoscale topographical cues on initial osteoblast morphology and migration. *Eur. Cell Mater.* **2010**, *20*, 329–343. [CrossRef] [PubMed]

66. Loesberg, W.A.; te Riet, J.; van Delft, F.C.; Schön, P.; Figdor, C.G.; Speller, S.; van Loon, J.J.W.A.; Walboomers, X.F.; Jansen, J.A. The threshold at which substrate nanogroove dimensions may influence fibroblast alignment and adhesion. *Biomaterials* **2007**, *28*, 3944–3951. [CrossRef] [PubMed]

67. Toworfe, G.K.; Composto, R.J.; Adams, C.S.; Shapiro, I.M.; Ducheyne, P. Fibronectin adsorption on surface-activated poly(dimethylsiloxane) and its effect on cellular function. *J. Biomed. Mater. Res. Part A* **2004**, *71*, 449–461. [CrossRef] [PubMed]

68. Lamers, E.; Walboomers, X.F.; Domanski, M.; te Riet, J.; van Delft, F.C.; Luttge, R.; Winnubst, L.A.J.A.; Gardeniers, H.J.G.E.; Jansen, J.A. The influence of nanoscale grooved substrates on osteoblast behavior and extracellular matrix deposition. *Biomaterials* **2010**, *31*, 3307–3316. [CrossRef] [PubMed]

69. Yim, E.K.F.; Darling, E.M.; Kulangara, K.; Guilak, F.; Leong, K.W. Biomaterials Nanotopography-induced changes in focal adhesions, cytoskeletal organization, and mechanical properties of human mesenchymal stem cells. *Biomaterials* **2010**, *31*, 1299–1306. [CrossRef] [PubMed]

70. Qi, L.; Li, N.; Huang, R.; Song, Q.; Wang, L.; Zhang, Q.; Su, R.; Kong, T.; Tang, M.; Cheng, G. The Effects of Topographical Patterns and Sizes on Neural Stem Cell Behavior. *PLoS ONE* **2013**, *8*, e59022. [CrossRef] [PubMed]

71. Sung, C.; Yang, C.; Yeh, J.A. Integrated Circuit-Based Biofabrication with Common Biomaterials for Probing Cellular Biomechanics. *Trends Biotechnol.* **2016**, *34*, 171–186. [CrossRef] [PubMed]

micromachines

MDPI

Article

Multiscale Analysis of Size Effect of Surface Pit Defect in Nanoindentation

Zhongli Zhang [1,2], Yushan Ni [1,*], Jinming Zhang [2], Can Wang [2] and Xuedi Ren [2]

[1]   Department of Aeronautics and Astronautics, Fudan University, Shanghai 200433, China;
      17110290016@fudan.edu.cn
[2]   Institute of Measurement and Testing Technology, Shanghai 201203, China;
      zhangjm@simt.com.cn (J.Z.); wangc@simt.com.cn (C.W.); renxd@simt.com.cn (X.R.)
*   Correspondence: niyushan@fudan.edu.cn; Tel.: +86-021-65642745

Received: 24 May 2018; Accepted: 10 June 2018; Published: 13 June 2018

check for
updates

Abstract: The nanoindentation on a pit surface has been simulated using the quasicontinuum method in order to investigate the size effect of surface pit defect on the yield load of thin film. Various widths and heights of surface pit defect have been taken into account. The size coefficient has been defined as an index to express the influence of the width or height of surface pit defect. The results show that as the size coefficient of width (of height) increases, at first the yield load of thin film decreases extremely slowly, until the size coefficient of width equals approximately one unit (half unit), at which point the yield load experiences an obvious drop. When the size coefficient of width (of height) reaches approximately two units (one unit), the yield load is almost the same as that of the nanoindentation on a stepped surface. In addition, the height of surface pit defect has more influence than the width on the yield load of thin film.

Keywords: multiscale; quasicontinuum method; surface pit defect; size effect

1. Introduction

As the development of a nanotechnique, nanoindentation [1] has already been a relatively simple and effective method for evaluating the material property of thin films. In order to get closer to the real system, a number of scientists have recently studied nanoindentation through simulations and experiments on thin film with defects such as inhomogeneities [2], grain boundaries [3,4], surface scratches [5], and surface steps [6,7]. Victor V. Pogorelko et al. [8] have found that nanohardness of coating is less than that of a single crystal Cu due to defects through their simulation. Telmo G. Santos et al. [9] have investigated how to identify micro- and nano-surface defects. Additionally, various kinds of surface defects have been artificially made to probe to friction and surface roughness phenomenon [10–12]. It is known to all that surface pit defect is very common in epitaxial thin films [13–15] and nanoimprint technology [16–18]. In fact, surface roughness has also been treated as part of the mixed group of surface pit defects [19]. Therefore, some scientific workers have focused on a pit surface to gain more information for the actual application of materials. Ni yushan et al. [20] studied nanoindentation of Al thin film compared with a surface defect situation and a defect free situation. The result shows that the initial surface defect has an obvious delay effect for the second dislocation emission, which indicates that a pit surface indeed plays an important role in the property of materials. Thus, it is necessary and significant to probe the influence of a pit surface in nanoindentation.

As the computer technology is highly developed, simulation methods, such as molecular dynamics (MD), become an efficient way to simulate nanoindentation experiments. However, the calculation accuracy and simulation efficiency of MD are limited by computer hardware, which make it impossible to use for large scale simulation. The quasicontinuum method (QC) is a multiscale method that combines continuum mechanics and molecular mechanics. It applies the MD model at the intense deformation region and a finite element model elsewhere so that the efficiency of the simulation is greatly improved to ensure accuracy.

It is difficult to study the nanoindentation on a pit surface through experiments. Thus far, relevant studies on this aspect are still rare to find. Our aim is to investigate the size effect of surface pit defect on the yield load in nanoindentation using the QC method, and hope it is an important directive to the defects testing or material application.

2. Methodology

The quasicontinuum method is applied in this simulation, which is brought by Tadmor [21]. It is one of the multiscale approaches that it keeps an atomistic description at highly deformed regions, whereas a linear elastic continuum method is implemented far away from this dislocation core. The QC method proceeds through molecular static energy minimization over an atomistic (non-local) domain and a finite element (local) domain. In this simulation, the Ercolessi–Adams potential (EAM) [22] is employed to describe the atomistic behavior of the system, and a finite element method is used at the linear deformation area of the material.

Figure 1 shows the nanoindentation model used in the simulation. The x-axis direction is direction, the y-axis direction is $[\bar{1}\,1\,0]$, and the outer-of-plane z direction is $[\bar{1}\,\bar{1}\,2]$ direction. This orientation is selected to facilitate dislocation emission. The width of Al thin film is 200 nm, and the height is 100 nm, which has one order of magnitude bigger than the usual MD level. The indenter is rigid with a width $4d_0$, where d_0 is 0.2328 nm (one atomic lattice spacing in [1 1 1] direction). The distance between the adjacent boundary of the indenter and surface pit defect is selected to be $4d_0$ (Figure 1), which is proved to be reasonable. When the surface pit defect is too close to the indenter, the deformation of surface pit defect is so severe that the crack is propagated under the tip of surface pit defect; when it is too far away from the indenter, the influence of pit defect on the yield load is too weak. According to the research on nanoindentation on a stepped surface [7], the spatial extent of the step's influence has been found out to be approximately three times the contact radius (measured mean stress at yield as a function of d/a_y; absolute value of distance from the step normalized by elastic contact radius prior to yield) on the surface, having step heights ranging from 5 to 30 Å. When $d/a_y < 2$, the yield stress is reduced by a neighboring step, while for $d/a_y > 3$, the yield stress is unaffected. The nanohardness are both calculated by the actual contact radius, though the indentation tip shape in this simulation and the nanoindentation experiment is square and round-like, respectively. Consequently, the change law of the nanohardness is reasonably similar if the ratio of surface defect distance to the actual contact radius is taken into account (just as the experiment discussed). In this simulation, the contact radius is $2d_0$ (half of the indenter width) and the distance between the left boundary of surface pit defect and the centre of the indenter is $6d_0$ ($4d_0 + 4d_0/2 = 6d_0$), which greatly agrees with this reference ($6d_0/2d_0=3$). Figure 1a,b are the simulation model of the width effect of surface pit defect. Ten different widths in this simulation are shown as D in Figure 1a, namely, $1d_0$, $2d_0$, $3d_0$, $4d_0$, $5d_0$, $6d_0$, $7d_0$, $8d_0$, $9d_0$, and $10d_0$, and the height of surface pit defect is a fixed height $5h_0$; Figure 1b shows the comparison model of surface step with height = $5h_0$; Figure 1c,d are the simulation model of the height effect of surface pit defect, where Figure 1c shows ten different heights H of surface pit defect, namely, $1h_0$, $2h_0$, $3h_0$, $4h_0$, $5h_0$, $6h_0$, $7h_0$, $8h_0$, $9h_0$, and $10h_0$, and the width of surface pit defect is a fixed value $5d_0$; Figure 1(d) shows the comparison model of surface step with height = $10h_0$. These parameters of width and height are selected in order to make a more comprehensive investigation. Further, in the out-of-plane direction, the thickness of this model is equal to the minimal repeat distance with the periodic boundary condition applied. The boundary condition of this single crystal Al thin film keeps it rigid at the bottom and free at the sides. The atoms under the indenter are forced to move gradually into the material by the displacement-imposed boundary condition. Each load step of the indenter is 0.02 nm, with a final depth 1.2 nm, because it is relatively proper and effective to catch the dislocation nucleation and mission with minimum total load steps. Because the width of the indenter is 1/200 of film width and the final depth is 1/100 of film height, it ensures that far-field boundary conditions do not affect the behavior in the vicinity of the indenter.

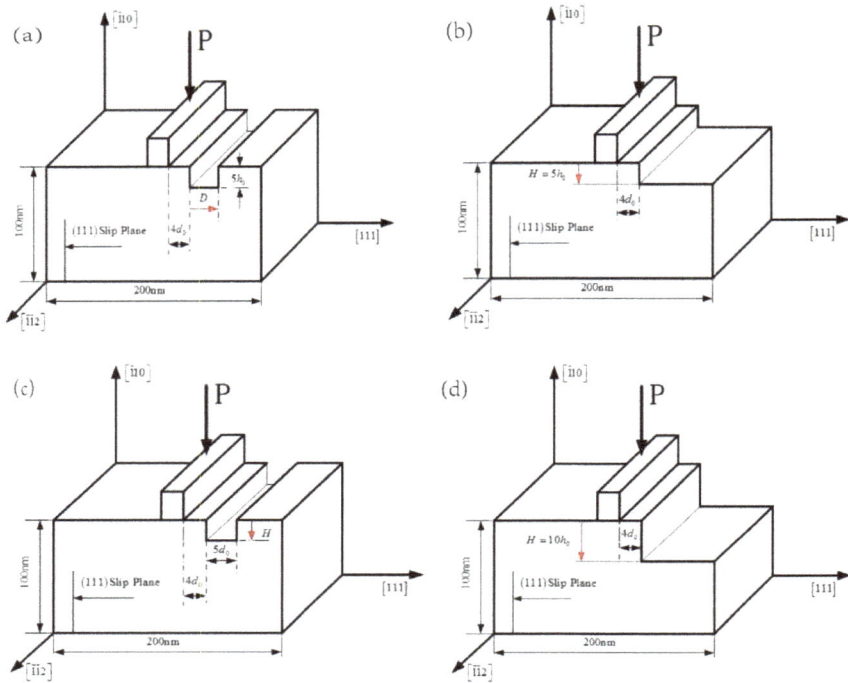

Figure 1. Schematic representation of the nanoindentation model of size effect: (**a**) width (D) changing from $1d_0$ to $10d_0$ of surface pit defect with the fixed height = $5h_0$; (**b**) the comparison model of surface step with height = $5h_0$; (**c**) height (H) changing from $1h_0$ to $10h_0$ of surface pit defect with the fixed width = $5d_0$; (**d**) the comparison model of surface step with height = $10h_0$.

The material of the model is single crystal Al thin film, and the crystallographic lattice constant a_1 is 0.4032 nm. One atomic spacing in [$\bar{1}$ 1 0] direction (h_0) is 0.1426 nm. Burgers vector $\vec{b}$ is 0.285 nm, shear modulus μ is 33.14 GPa, Poisson ν is 0.319, and (1 1 1) surface energy γ_{111} is 0.869 J/m². Figure 2 shows the schematic of local and non-local representative atoms and tessellation during nanoindentation on ($\bar{1}$ $\bar{1}$ 2) plane of Al film with initial surface pit defect, where the red square is the rigid indenter and the blue filled circles are the non-local representative atoms, while the green ones are the local representative atoms. The system investigated here is very large by current atomistic modeling standards. A standard lattice statics analysis for this system would treat millions of atoms and would have to be performed on a parallel supercomputer. By using the quasicontinuum method, the computational intensity is greatly reduced. Regarding this single crystal Al system with a size of 100 nm × 200 nm, only 5000 atoms are treated explicitly at most (15,000 degrees of freedom), and a simulation can be finished on a common personal computer in a few days.

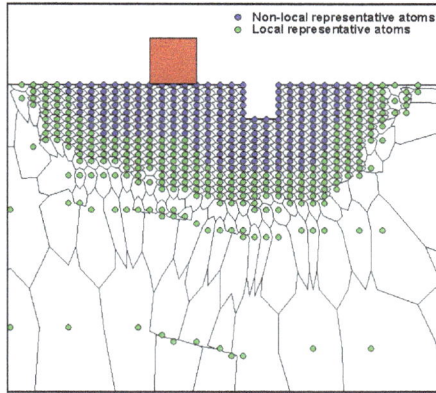

Figure 2. Schematic of local and non-local representative atoms with initial surface pit defect.

3. Results

3.1. Width Effect of Surface Pit Defect on Yield Load

It has long been recognized that the yield load of materials is one of the most important indexes of material properties. At the load-displacement curve, the yield load corresponds to the first highest point at the initial linear portion, which indicates onset of the dislocation emission. Further, the yield load of materials can be obviously influenced by defects such as surface pit defect. In the present paper, ten different widths of surface pit defect are simulated, from $D = 1d_0$ to $10d_0$, in order to probe the width effect of surface pit defect on yield load. Figure 3 shows the yield load curve as the width of surface pit defect changes. It can be found out that the yield load of thin film with surface pit defect generally displays a tendency to decrease, which is reasonable because the structure of thin film is destroyed more and more severely by the increase of the width of surface pit defect. When the simulated width increases from $D = 1d_0$ to $7d_0$, the yield load decreases extremely slowly; after the width reaches $7d_0$, the yield load experiences an obvious drop from 14.8 N/m to 14.24 N/m. Then, the yield load curve displays the phenomenon of a slow decrease again.

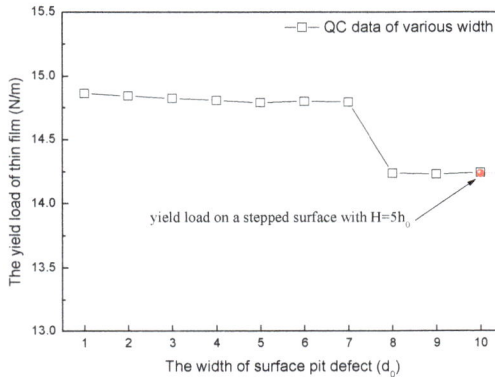

Figure 3. The yield load of thin film as the width changing of surface pit defect (with a standard deviation of 0.01 N/m). QC—quasicontinuum method.

In order to conduct a comprehensive investigation of the width effect of surface pit defect, the nanoindentation on a stepped surface has been carried out for comparison (namely, the simulation width of surface pit defect is infinitely large), as shown in Figure 1b. The results show that the yield load of nanoindentation on a stepped surface with $H = 5h_0$ is approximately 14.23 N/m, which is very close to the yield load value of $D = 10d_0$ (the red point in Figure 3). That is to say, when the width of surface pit defect increases to $10d_0$, the yield load of thin film almost reaches the yield load value of nanoindentation on a stepped surface.

3.2. Height Effect of Surface Pit Defect on Yield Load

An investigation of the height effect of surface pit defect on yield load has also been carried out. Ten different heights of surface pit defect are simulated, from $H = 1h_0$ to $10h_0$, with a fixed width $D = 5d_0$ (as shown in Figure 1c). Figure 4 shows the yield load curve as the height of surface pit defect changes. It can be found out that the change law of the yield load of thin film is very similar to the situation of the width effect. As the simulation height increases from $H = 1h_0$ to $5h_0$, the yield load decreases extremely slowly, until the height reaches $6h_0$, at which point the yield load experiences an obvious drop from 14.79 N/m to 14.14 N/m. Then, the yield load curve slowly decreases again.

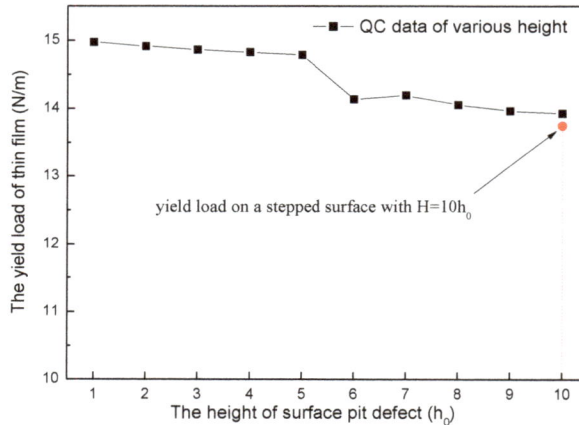

Figure 4. The yield load of thin film as the height changing of surface pit defect (with a standard deviation of 0.01 N/m).

The nanoindentation on a stepped surface with the $10h_0$ step height has been investigated for comparison, as shown in Figure 1d. The results show that the yield load of nanoindentation on such a stepped surface is approximately 13.75 N/m (the red point in Figure 4). It can be easily found out that when the simulation height of surface pit defect increases to $10h_0$, the yield load of thin film is about 13.93 N/m, which is already close to the yield load of nanoindentation on a stepped surface.

4. Discussion

4.1. The Investigation of Dislocation Nucleation and the Estimation of Peierls Stress

In order to probe the reason for such an obvious decline of yield load ($D = 7d_0$ to $8d_0$ section in Figure 3, $H = 5h_0$ to $6h_0$ section in Figure 4), relevant snapshot of atoms under the indenter and corresponding out-of-plane displacement plot are probed. The results show that when the thin film yields, two dissociated <1 1 0> edge dislocations are emitted beneath the indenter after nucleation. Considering there are too many snapshots, the situation of $D = 1d_0$ in width effect simulation and

H = 1h$_0$ in height effect simulation are carried out for example. The dislocated structure beneath the indenter is given in Figure 5, along with the out-of-plane displacements experienced by the atoms, where dimensions and displacements are in 0.1 nm. The nucleated dislocations are easily seen through UZ contours displayed in Figure 5. The out-of-plane displacements in the stacking fault regions between the partials are a clear fingerprint of the location of the dislocations. The repeat distance in the out-of-plane direction of the crystal structure is 0.4938 nm for this model. It can be found out that the dislocations are composed of 1/6 <1 1 2> Shockley partials that bound a stacking fault. On the left,

$$\frac{1}{2}[\overline{1}10] = \underbrace{\frac{1}{6}[\overline{1}2\overline{1}]}_{top} + \underbrace{\frac{1}{6}[\overline{2}11]}_{bottom} \tag{1}$$

and on the right,

$$\frac{1}{2}[1\overline{1}0] = \underbrace{\frac{1}{6}[1\overline{2}1]}_{top} + \underbrace{\frac{1}{6}[2\overline{1}\overline{1}]}_{bottom} \tag{2}$$

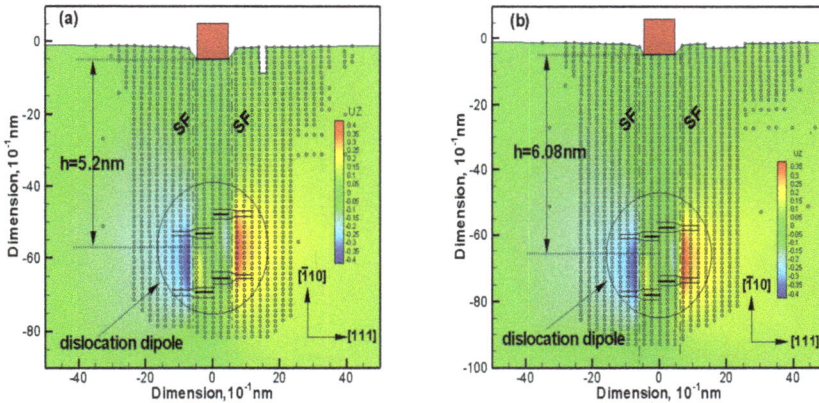

Figure 5. Snapshot of atoms under the indenter and corresponding out-of-plane displacement plot, where UZ is atom displacement at out-of-plane: (**a**) width changing D = 1d$_0$ at the yield of thin film; (**b**) height changing H = 1h$_0$ at the yield of thin film.

In Figure 5a, the dislocation dipole travels into bulk after nucleation at the load step of 0.5 nm, and its centre settles at the depth of 5.2 nm. In Figure 5b, the dislocation dipole travels into bulk at the same load step of 0.5 nm, but its centre settles at the depth of 6.08 nm. Further, when compared with all these snapshots of atom structures in the simulation of size effect, it is found out that when the size of surface pit defect changes, there is a different emission depth of dislocations (see Figure 5a,b, for example). That is to say, most likely the different yield load of thin film in macroscopy corresponds to the emission depth of dislocation in microscopy.

For the purpose of the explanation, such change law of the yield load, these emission depths of dislocations are used as an equilibrium distance to further obtain an estimate for the Peierls stress predicted by the EAM potential [22]. Because Peierls stress is actually the resisting force during the dislocation movement resulting from the lattice structure, the change of the yield load can be reasonably explained by Peierls stress. Aside from the lattice friction, there are two forces acting on the dislocation: (i) the Peach–Koehler force (F_{PK}) due to the indenter stress field driving the dislocation into bulk; (ii) the image force (F_I) pulling the dislocation to the surface. The force on the dislocation is the sum of these two forces. The dislocation escapes the attractive region and propagates into the bulk, and is finally stopped by lattice friction. That is to say, the force on the dislocation will be balanced at the equilibrium depth by the lattice friction force that results from the Peierls stress (σ_p) [23].

$$F_{PK} + F_1 = b\sigma_p \tag{3}$$

To compute the Peach–Koehler force, shear stress field is required beneath the indenter. In this simulation, there is a frictionless rectangular indenter acting on an elastic body occupying the lower half-plane, $y < 0$, the shear stress in bipolar coordinates is [24]

$$\sigma_{xy} = -\frac{Pr^2\sin\theta}{\pi(r_1 r_2)^{3/2}} \sin[\theta - \frac{3}{2}(\theta_1 + \theta_2)] \tag{4}$$

where P is the indentation load. According to the coordinate system of $2a$ indentation contact (the width of indenter is $2a$), as shown in Figure 6, at a depth h beneath the right indenter tip, there is $r = \sqrt{a^2 + h^2}$, $r_1 = h$, $r_2 = \sqrt{4a^2 + h^2}$, $\theta = -\tan^{-1}h/a$, $\theta_1 = -\pi/2$, $\theta_2 = -\tan^{-1}(h/2a)$. The resulting Peach–Koehler force is

$$F_{PK}(h) = (\mathbf{b} \cdot \boldsymbol{\sigma}) \times \boldsymbol{\ell} = b\sigma_{xy}(h) \tag{5}$$

where b is the Burgers vector, σ is the applied stress tensor, and ℓ is the dislocation line vector.

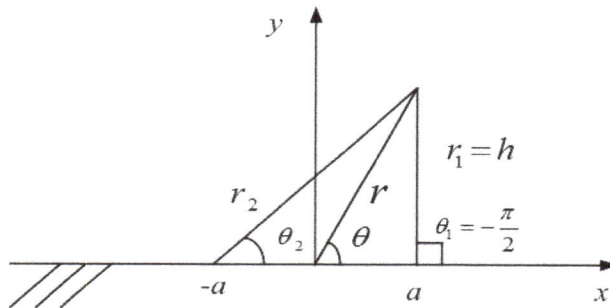

Figure 6. Bipolar coordinate for a $2a$ indentation contact.

The image force acting on one of the dislocations of a dipole of width $d = 2a$ at depth h beneath the indenter can be shown to be

$$F = \frac{\mu b^2}{\pi(1-v)}[\frac{1}{4h} - \frac{4h^3(4h^2 - 3d^2)}{(4h^2 + d^2)^3}] \tag{6}$$

According to the discussion above, Peierls stress in every size of surface pit defect has been calculated and plotted. Figure 7 shows the variation of Peierls stress in the simulation of width effect. It can be easily found out that when the width of surface pit defect changes from D = $1d_0$ to $7d_0$, the Peierls stress fluctuates narrowly at the value of 100 MPa. When the width increases to more than $8d_0$, the Peierls stress abruptly obviously drops down to about 70 MPa. Such change law is greatly in keeping with the variation of yield load in the width effect simulation. In a similar manner, it can be also found out from Figure 8 that when the height of surface pit defect changes from H = $1h_0$ to $5h_0$, the Peierls stress fluctuates narrowly at the value of 70 MPa. When the height increases to more than $6h_0$, the Peierls stress abruptly obviously drops down to about 50 MPa, which is also in accordance with the variation of yield load in the height effect simulation. That is to say, such an obvious decline of yield load (D = $7d_0$ to $8d_0$ section in Figure 3, H = $5h_0$ to $6h_0$ section in Figure 4) results from the severe reduction of the Peierls stress, which is caused by the size increase of surface pit defect. Consequently, it is reasonable and useful to explain the variation of yield load through the Peierls stress.

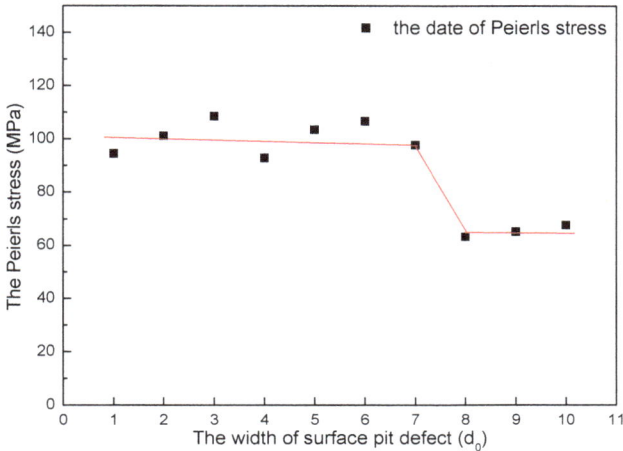

Figure 7. The variation of Peierls stress in the simulation of width effect (with a standard deviation of 0.2 MPa).

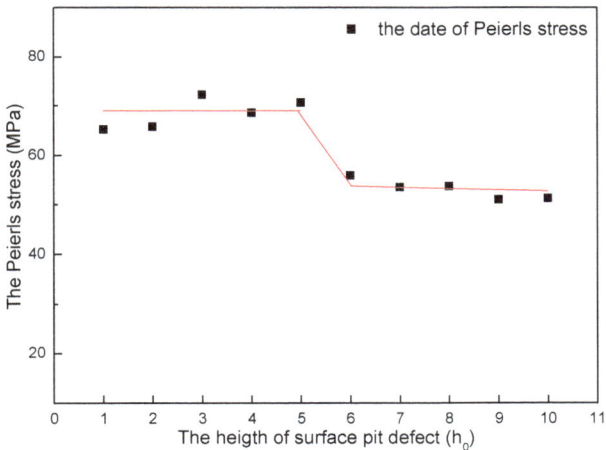

Figure 8. The variation of Peierls stress in the simulation of height effect (with a standard deviation of 0.1 MPa).

4.2. Size Coefficient

It can be figured out that the turning point ($D = 7d_0$ in the width effect simulation while $H = 5h_0$ in the height effect simulation) is different in this simulation. That is to say, the influence degree of width parameter is different from the height parameter of surface pit defect. Thus, a further discussion is carried out to quantify the size effect of surface pit defect. It is reasonable that the influence on the hardness and yield load of thin film would be much more severe if the surface pit defect gets closer to the indenter. That is to say, if the same degree of hardness damage is made by surface pit defect, the larger size of pit defect is needed where it is farther away from the indenter. Consequently, in order to define a more precise expression of the size effect of surface pit defect, a size coefficient α should be carried out as follows:

$$\alpha = \frac{L^*}{d^*} \tag{7}$$

where "L^*" means the characteristic length of surface pit defect (namely the width D in the width effect simulation and the height H in the height effect simulation), and "d^*" means the distance between the center of the indenter and the left boundary of the surface pit defect (in this simulation, d^* is a constant $6d_0$).

In the width effect simulation, the critical width of an abrupt obvious drop of yield load is $7d_0$ (at the point D = $7d_0$ in Figure 3). Thus, the size coefficient α is approximately 1.17 ($\frac{L^*}{d^*} = \frac{D}{d^*} = \frac{7d_0}{6d_0} = \frac{7}{6}$). When α reaches approximately 2 ($\frac{L^*}{d^*} = \frac{D}{d^*} = \frac{10d_0}{6d_0} = 1.7$), as shown in Figure 3 at the point D = $10d_0$, the yield load of thin film is almost the same with that of nanoindentation on a stepped surface (the red point in Figure 3).

In the height effect simulation, the critical height of an abrupt obvious drop of yield load is $5h_0$ (at the point H = $5h_0$ in Figure 4). Then, the size coefficient α is approximately 0.51 ($\frac{L^*}{d^*} = \frac{H}{d^*} = \frac{5h_0}{6d_0} = 0.51$). When α reaches approximately 1 ($\frac{L^*}{d^*} = \frac{H}{d^*} = \frac{10h_0}{6d_0} = 1.02$), as shown in Figure 4 at the point H = $10h_0$, the yield load of thin film is almost the same as that of nanoindentation on a stepped surface (the red point in Figure 4).

It can be found out that the size coefficient of height is almost half of the size coefficient of width in the abrupt obvious drop point of yield load decline, which suggests that the height parameter of surface pit defect plays a more important role than width parameter.

In addition, from the point of the area of surface pit defect, it also can be proved that the height of surface pit defect is a leading factor on yield load. Figure 9 shows the yield load of thin film changing as the area changes. It can be easily found out that the slope of yield load curve through the increase of height is bigger than the one through the increase of width. It indicates that the increase of height makes the yield load decrease faster. When the area of surface pit defect increases from $5h_0d_0$ to $25h_0d_0$, the yield load through the increase of width is smaller than the one through the increase of height. This is because during this internal area, the height value of surface pit defect in the curve of height increase (red curve in Figure 9) is bigger than the other one (black curve). However, when the area is larger than $25h_0d_0$, the yield load through the increase of height is smaller than the one through the increase of width. This is because the height of surface pit defect in the curve of height increase is over $6h_0$, while the height of surface pit defect in the curve of width increase is still $5h_0$. According to the discussion above, the height of surface pit defect makes more influence than width in the yield load of thin film, which indicates that the height of the pit is a leading factor on the influence of the yield load in nanoindentation.

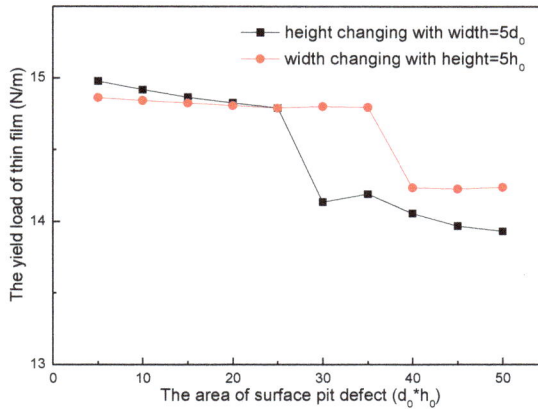

Figure 9. The yield load of thin film as the area changing of surface pit defect.

5. Conclusions

In this paper, the QC method is employed to investigate the size effect of surface pit defect on yield load in nanoindentation. The conclusion can be drawn as follows:

- As the width of surface pit defect increases, the yield load of thin film decreases extremely slowly, until the size coefficient of width equals approximately one unit, at which point the yield load experiences an obvious drop. When the size coefficient of width reaches approximately two units, the yield load is almost the same as that of the nanoindentation on a stepped surface.
- As the height of surface pit defect increases, the yield load of thin film decreases extremely slowly, until the size coefficient of height equals approximately half unit, at which point the yield load experiences an obvious drop. When the size coefficient of height reaches one unit, the yield load is almost the same as that of the nanoindentation on a stepped surface.
- The height of surface pit defect has more influence than the width on the yield load of thin film, which suggests that the height of the pit is a leading factor on the influence of yield load. Such investigation results in this simulation may have important directive to the defects testing or material application.

Based on such a size effect of surface defect in nanohardness in the present paper, a further work of surface defect effect might be interesting and worth focusing on if the surface defect is not a cavity but another material, which is usually seen in alloy.

Author Contributions: Data curation, Z.Z., J.Z., C.W. and X.R.; Formal analysis, Z.Z.; Funding acquisition, Y.N.; Investigation, Z.Z.; Methodology, Y.N.; Project administration, Y.N.; Supervision, Y.N.; Validation, Z.Z.; Visualization, Z.Z., J.Z., C.W. and X.R.; Writing—original draft, C.W.; Writing—review & editing, Z.Z.

Acknowledgments: This work is supported by the National Natural Science Foundation of China (Grant No. 11572090). The authors would like to thank Tadmor E. B. and Miller R. for their quasicontinuum code and suggestions during the multiscale simulations.

References

1. Oliver, W.C.; Pharr, G.M. An improved technique for determining hardness and elastic modulus using load and displacement sensing indentation experiments. *J. Mater. Res.* **1992**, *7*, 1564–1583. [CrossRef]

2. Mitchell, J.W. *Growth and Perfection of Crystals*; Doremus, R.H., Roberts, B.W., Turnbull, D., Eds.; Wiley: New York, NY, USA, 1958; pp. 386–389.

3. Yang, B.; Vehoff, H. Dependence of nanohardness upon indentation size and grain size—A local examination of the interaction between dislocations and grain boundaries. *Acta Mater.* **2007**, *55*, 849–856. [CrossRef]

4. Soifer, Y.M.; Verdyan, A.; Kazakevich, M.; Rabkin, E. Nanohardness of copper in the vicinity of grain boundaries. *Scr. Mater.* **2002**, *47*, 799–804. [CrossRef]

5. Griffith, A.A., VI. The phenomena of rupture and flow in solids. *Philos. Trans. R. Soc. Lond. A* **1921**, *221*, 163. [CrossRef]

6. Shan, D.; Yuan, L.; Guo, B. Multiscale simulation of surface step effects on nanoindentation. *Mater. Sci. Eng. A* **2005**, *412*, 264–270. [CrossRef]

7. Kiey, J.D.; Hwang, R.Q.; Houston, J.E. Effect of Surface Steps on the Plastic Threshold in Nanoindentation. *Phys. Rev. Lett.* **1998**, *81*, 4424–4427.

8. Pogorelko, V.V.; Krasnikov, V.S.; Mayer, A.E. High-speed collision of copper nanoparticles with aluminum surface: Inclined impact, interaction with roughness and multiple impact. *Comput. Mater. Sci.* **2018**, *142*, 108–121. [CrossRef]

9. Santos, T.G.; Inácio, P.L.; Costa, A.A.; Miranda, R.M.; de Carvalho, C.C.C.R. Applications of a new NDT technique based on bacterial cells. *NDT E Int.* **2016**, *78*, 20–28. [CrossRef]

10. Xu, F.L.; Xin, Y.S.; Li, T.S. Friction-induced surface textures of liquid crystalline polymer evaluated by atomic force microscopy, spectroscopy and nanoindentation. *Polym. Test.* **2018**, *68*, 146–152. [CrossRef]

11. Erinosho, M.F.; Akinlabi, E.T.; Johnson, O.T. Characterization of surface roughness of laser deposited titanium alloy and copper using AFM. *Appl. Surf. Sci.* **2018**, *435*, 393–397. [CrossRef]

12. Pei, H.Q.; Wen, Z.X.; Li, Z.W.; Zhang, Y.M.; Yue, Z.F. Influence of surface roughness on the oxidation behavior of a Ni-4.0Cr-5.7Al single crystal superalloy. *Appl. Surf. Sci.* **2018**, *440*, 790–803. [CrossRef]

13. Tromas, C.; Stinville, J.C.; Templier, C.; Villechaise, P. Hardness and elastic modulus gradients in plasma-nitrided 316L polycrystalline stainless steel investigated by nanoindentation tomography. *Acta Mater.* **2012**, *60*, 1965–1973. [CrossRef]

14. Mao, W.G.; Shen, Y.G.; Lu, C. Nanoscale elastic–plastic deformation and stress distributions of the C plane of sapphire single crystal during nanoindentation. *J. Eur. Ceram. Soc.* **2011**, *31*, 1865–1871. [CrossRef]

15. Chang, T.R.; Tsai, C.H. Mechanical responses of Zn1-xMnxO epitaxial thin films. *Appl. Surf. Sci.* **2011**, *258*, 614–617. [CrossRef]

16. Li, K.; Morton, K.; Veres, T.; Cui, B. Nanoimprint Lithography and Its Application in Tissue Engineering and Biosensing. *Compr. Biotechnol.* **2011**, *5*, 125–139.

17. Park, S.Y.; Choi, K.B.; Lim, H.J.; Lee, J.J. Fabrication of a nano-scale embedded metal electrode in flexible films by UV/thermal nanoimprint lithography tools. *Microelectron. Eng.* **2011**, *88*, 1606–1609. [CrossRef]

18. Taylor, H.; Smistrup, K.; Boning, D. Modeling and simulation of stamp deflections in nanoimprint lithography: Exploiting backside grooves to enhance residual layer thickness uniformity. *Microelectron. Eng.* **2011**, *88*, 2154–2157. [CrossRef]

19. Jiang, W.G.; Su, J.J.; Feng, X.Q. Effect of surface roughness on nanoindentation test of thin films. *Eng. Fract. Mech.* **2008**, *75*, 4965–4972. [CrossRef]

20. Li, J.W.; Ni, Y.S.; Lin, Y.H.; Luo, C. Multiscale simulation of nanoindentation on Al thin film. *Acta Metall. Sin.* **2009**, *45*, 129–136.

21. Tadmor, E.B. The Quasicontinuum Method. Ph.D. Thesis, Brown University, Providence, RI, USA, 1996.

22. Daw, M.S.; Baskes, M.I. Semiempirical, Quantum Mechanical Calculation of Hydrogen Embrittlement in Metals. *Phys. Rev. Lett.* **1983**, *50*, 1285. [CrossRef]

23. Tadmor, E.B.; Miller, R.; Phillips, R.; Ortiz, M. Nanoindentation and incipient plasticity. *J. Mater. Res.* **1999**, *14*, 2233–2250. [CrossRef]

24. Muskhelishvili, N.I. *Some Basic Problems of the Mathematical Theory of Elasticity*, 3rd ed.; P. Noordhoff Ltd.: Groningen, The Netherlands, 1953; pp. 481–483.

micromachines

MDPI

Review
Nanoindentation of Soft Biological Materials

Long Qian and Hongwei Zhao *

School of Mechanical and Aerospace Engineering, Jilin University, Changchun 130025, China;
qianlong17@mails.jlu.edu.cn
* Correspondence: hwzhao@jlu.edu.cn; Tel.: +86-0431-85095757

Received: 29 October 2018; Accepted: 5 December 2018; Published: 11 December 2018

check for
updates

Abstract: Nanoindentation techniques, with high spatial resolution and force sensitivity, have recently been moved into the center of the spotlight for measuring the mechanical properties of biomaterials, especially bridging the scales from the molecular via the cellular and tissue all the way to the organ level, whereas characterizing soft biomaterials, especially down to biomolecules, is fraught with more pitfalls compared with the hard biomaterials. In this review we detail the constitutive behavior of soft biomaterials under nanoindentation (including AFM) and present the characteristics of experimental aspects in detail, such as the adaption of instrumentation and indentation response of soft biomaterials. We further show some applications, and discuss the challenges and perspectives related to nanoindentation of soft biomaterials, a technique that can pinpoint the mechanical properties of soft biomaterials for the scale-span is far-reaching for understanding biomechanics and mechanobiology.

Keywords: nanoindentation; mechanical properties; soft biomaterials; viscoelasticity; atomic force microscopy (AFM)

1. Introduction

Mechanical behavior of biological materials has come to the front stage recently, not only since its importance from the mechanical and load-bearing viewpoints, but also in the way that it influences other bio-functionalities [1]. Recent studies have directly linked major biological performances, mechanisms, and diseases to the mechanical response from the biomolecular up to the organ level [2–7]. In addition to the many medical applications, mechanical characterization of biological materials has also fueled the recent growth of materials science and engineering applications—bionics [8], whereas the mechanical characterization of soft biomaterials, especially down to biomolecules, is more difficult and fraught with more pitfalls compared with the hard biomaterials.

At present, aiming at characterizing soft biomaterials, a variety of testing techniques have been developed and utilized widespread from bulk scale to the micro/nano-scale [9–14]. While every technique has its pros and cons, nanoindentation (including AFM) is considered as a powerful tool to conduct mechanical analyses, especially down to micro/nano-scale with nanometer depth and sub-nanonewton force resolution [15]. Firstly, and the most attractively, its micro/nano-mechanical contact methodology allow for region-specific mapping of biomaterials inhomogeneity [16,17], and studying of cell mechanics [18]. Secondly, its unrestriction of tissue morphology avoids special preparation of material samples, which can be taken advantage of in the field of in vivo testing [19,20]. Thirdly, its capacity of coupling with other optical techniques provides a new horizon for mechanical characterization [21,22]. Fourthly, compared with traditional testing methods with a single-measurement mode, indentation can provide an ideal loading modality whose stress and strain fields comprise tension, compression, and shear loading modes [23]. Further, as a non/micro-invasive method, nanoindentation can also be utilized for some valuable samples, such as fossils that are millions of years old [24].

In regard to nanoindentation of soft biological materials, current measurements lack a standardized testing routine owing to several challenges we are facing:

1. Soft biomaterials are significantly less stiff than typical engineering materials, which may be to some extent in conflict with the range/resolution of some commercial instruments.
2. Most nanoindentation instrumentations are designed based on "dry state" testing, lacking environmentally-controlled components to build physiological environment (such as fluid submersion and thermal control).
3. For most commercial instruments with the optical microscopes, the interference between the indenter/microscope and the sample surface may exist in the switch (movement) of indenter/microscope owing to non-ideal surface. This may lead to a considerable challenge, particularly in in vitro and in vivo testing.
4. There is a paucity of consensus on the appropriate data analysis to mechanical characterization of soft biomaterials.

In addition to the above factors, some features of soft biomaterials, such as viscoelasticity and adhesion, may also make some deviations when testing. These challenges have limited the development and utilization of nanoindentation technique which, in turn, slow the footsteps of characterization of soft biomaterials. In this review, we will discuss the above factors, and conclude with the perspectives and opportunities of nanoindentation of soft biomaterials.

2. Soft Biomaterials

For most people, soft materials are materials where the deformation can be felt by hand or seen with the naked eye without applying an excessive force [25], which reflects the feature in a straightforward manner: it is much more compliant than many engineering materials. In fact, the words "soft" and "hard" do not indicate anything in regard to hardness or plastic deformation exactly, and soft biomaterials only imply non-mineralised in their healthy state [26]. Soft biomaterials, such as globular proteins [27], cancer cells [28], arteries [29], cartilage [30], and the brain [31], vary in multiple length scales from the molecular via the cellular and tissue all the way to the organ level, and the above complex hierarchical structures make characterization within wide physiologically-relevant timescales and the elastic modulus range (Figure 1).

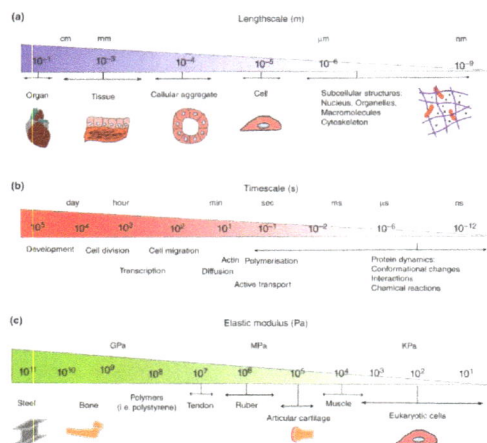

Figure 1. Multiple scales of soft biomaterials. (**a**) Length scales from the molecular to the organ level; (**b**) timescales of different physiological processes; and (**c**) comparisons of the elastic modulus among different typical materials. Reproduced with permission from [32].

A number of constitutive models, such as linear elastic, hyperelastic, viscoelastic, and poroelastic models, have been widely used to mechanical characterization of soft biomaterials. In fact, these models can fall into different aspects of the constitutive response, in particular that distinguish them from typical engineering materials.

First, many soft biomaterials exhibit nonlinear stress–strain ($\sigma - \varepsilon$) behavior, with an increasing stiffness as the strain increases. Nonetheless, the stress can be considered proportional to the strain under small deformations (Figure 2a), in which the relationship between stress and strain is governed by a constant elastic modulus E (just as the elastic stage of typical engineering materials), and can be fitted to linear elastic model ($\sigma = E\varepsilon$)—the simplest type of material response. The threshold of the linear range varies from different biomaterials and testing methods [33], and further discussion of the linear level can be seen in Section 4.3. Another strategy of simplification of nonlinear behavior with the linear elastic model is to divide the stress profile into several portions. At each portion, the mechanical behavior can be considered linear and characterized with the linear elastic model [34].

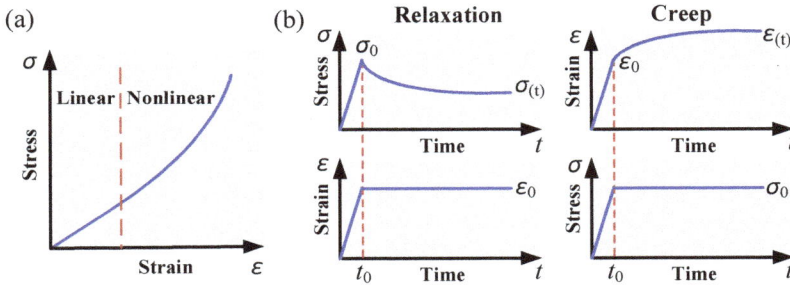

Figure 2. Constitutive responses of soft biomaterials. (**a**) Nonlinear stress–strain ($\sigma - \varepsilon$) behavior, in which the stress profile can be considered linear under small strain; and (**b**) time-dependent mechanical behavior: stress-relaxation (**left**) and creep (**right**).

To acquire more accurate constitutive data, hyperelasticity provides a means of modeling the nonlinear stress–strain behavior of soft biomaterials. Many different kinds of hyperelastic models could be used to characterize mechanical behavior, such as Neo–Hookean, Mooney–Rivlin, Ogden, and Yeoh models [35,36]. As far as the above models are concerned, the Ogden model is classically used for finite element simulations, and Neo–Hookean, Mooney–Rivlin, and Yeoh models are best used at low, moderate, and high strain levels, respectively [37]. More details about different hyperelastic models and their parameters determined from indentation can be found in the literature [37,38].

The second aspect of soft biomaterials is the time-dependent mechanical behavior, i.e., viscosity. Viscoelastic model, displaying a combination of both elastic and time-dependent responses. A viscoelastic material stores and dissipates mechanical energy simultaneously undergoing imposed mechanical excitation, with the response of stress-relaxation or creep over time, as shown in Figure 2b. Typically, the model formulated by a Prony series approximation can be used to describe viscoelastic response:

$$G(t) = G_\infty + \sum_j G_j \cdot e^{-t/\tau_j} \tag{1}$$

where G_∞ is the equilibrium shear modulus, τ_j is the time constant for each exponential term, and G_j is the associated magnitude of shear modulus. The initial shear modulus can be calculated by summing G_∞ and G_j. The above parameters can be predicted using a minimization algorithm [39]. Additionally, nonlinear viscoelasticity can also been modeled during indentation [40].

As for most soft biomaterials, they are also hydrated, and a poroelastic (also called biphasic) model can also be used to describe time-dependent behavior under physiological condition: the flow of a fluid through a porous elastic solid [41]. In addition to the shear modulus G and the (drained)

Poisson's ratio v used to characterize the elastic behavior of the porous skeleton, Darcy (hydraulic) permeability κ, formulated between the fluid viscosity η and the intrinsic permeability k ($k = \kappa\eta$), is an important parameter to indicate the fluid–solid coupling and the fluid flow [42]. More details about poroelasticity during nanoindentation of soft biomaterials will be discussed in the Section 4.5.

Another aspect of soft biomaterials is the anisotropy and the heterogeneity. Some soft biomaterials exhibit direction-dependent and region-specific behavior based on their structural complexities at multiple length scales [17,43,44]. Apart from inherent complexities of materials, some external conditions may also lead to the above behavior [45,46]. It is imperative to characterize this structure-mechanical behavior with some specific analysis methods and models [46–48].

3. Methods

Two kinds of commercial instruments have been developed and utilized widespread for indentation testing, especially down to micro/nano-scale: dedicated instrumented indentation instruments (nanoindenter) and the atomic force microscope (AFM). In this section, we will introduce the working principles of two different instruments, and discuss the analysis of experimental data based on the force-indentation curves.

3.1. Nanoindenter

Nanoindentation, a form of depth-sensing indentation (DSI) testing technique, involves the application of a controlled load/depth to the surface to induce local surface deformation (Figure 3). During a typical testing, Load P, indentation depth h and time t are monitored as the indenter is actuated into the test material's surface. The response of $P - h - t$ trace is fitted to a range of different constitutive models to identify mechanical properties of the sample.

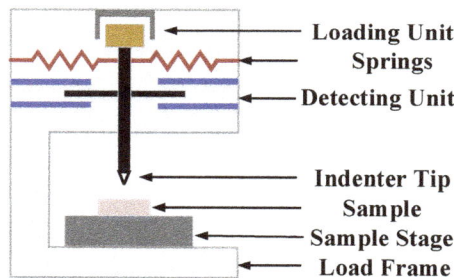

Figure 3. Schematic diagram of a nanoindenter instrument.

Usually, a nanoindenter consists of several essential components:

- Loading unit: typically actuated by the expansion of the piezoelectric element, magnetic coils, or electrostatically [49].
- Detecting unit: sensors (capacitance or inductance) to record the displacement of the indenter. In fact, whether the strategy applying force and measuring displacement through separate means, or using the same transducer, the data of raw force and displacement are always coupled due to the leaf springs [50].
- Indenter tip: for soft biomaterials, typically using dull indenters (such as spherical and flat-ended), rather than sharp indenters (such as Berkovich and Vickers) to avoid penetration of the sample.
- Sample stage: a two- or three-coordinate stage (x, y, or z) to move the sample.
- Microscopes: to observe and choose the point of the sample during testing.
- Control system: the computer with the software to operate instrument, analyze results, and save data.

- Other additive components: such as custom irrigation system [51] and fluid cell [52] for biomaterials.

Definitely, the design of instruments may be not identical to the above components, especially for the custom instruments, but part of them and similar working principles are, at least, included [31,53].

3.2. Atomic Force Microscopy (AFM)

Atomic force microscopy (AFM), a part of the scanning probe microscopy branch, operates based on the interaction between the sample surface and the small tip located on the end of a sensitive cantilever. Apart from the basic goal of imaging surface morphology, AFM can also act as a powerful instrument to conduct micro/nano-mechanical analysis. In particular, the mode of mechanical characterization combined with high resolution imaging technique, allows a more targeted investigation of biomaterials features, especially down to nanoscale.

The setup and the measurement of AFM have been covered in a number of works [54–56], and a brief introduction is given here. During typical testing, the tip or the sample is moved to each other with a piezoelectric, until the tip-sample contact occurs and the cantilever deflects (Figure 4). The indentation force P can be described by Hooke's law:

$$P = k_c \Delta d \tag{2}$$

where k_c denotes the spring constant of the cantilever, and Δd is the corresponding deflection of the cantilever.

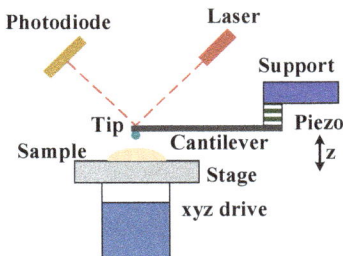

Figure 4. Schematic diagram of an AFM instrument.

For stiff samples, which are several orders of magnitude stiffer than the tip, the displacement of the piezo Δz is equal to corresponding deflection of the cantilever Δd ($\Delta z = \Delta d$), since the indentation depth h is zero ($h = 0$). However, for many samples (e.g., soft biomaterials), the displacement of the piezo Δz is larger than the corresponding deflection of the cantilever Δd owing to the indentation, and the indentation depth h can be expressed as following:

$$h = \Delta z - \Delta d \tag{3}$$

So far, the "deflection–displacement curve" of AFM indentation can be developed to typical "force-indentation ($P - h$) curve" in nanoindentation, and analyzed likewise by different constitutive models.

3.3. Force–Indentation ($P - h$) Curves

3.3.1. Oliver–Pharr Model

The analysis of force–indentation ($P - h$) curves of commercial nanoindnetation system is often based on the work by Doerner and Nix [57] and Oliver and Pharr [58]. It is assumed that only the elastic recovery occurs in the initial portion of the unloading, while the loading is an elastic–plastic contact mode. The relationship formulated between indentation force P and depth h during unloading can be expressed a power law relation:

$$P = \alpha(h - h_f)^m \tag{4}$$

where h_f is the final residual indent of depth, α and m are power law fitting constants related to the indenter geometry. The stiffness S, defined as the resistance in response to an applied force, can be calculated by the slope of the upper portion of the unloading curve:

$$S = \frac{dP}{dh} \tag{5}$$

Further, the reduced elastic modulus E_r (the combined modulus of the tip and the sample) can be determined in the terms of the unloading the contact area A:

$$E_r = \frac{\sqrt{\pi}}{2} \frac{S}{\sqrt{A}} \tag{6}$$

and the sample elastic modulus E_s can be determined by decoupling the deformations of both indenter and the sample as given by:

$$\frac{1}{E_r} = \frac{(1 - v_s{}^2)}{E_s} + \frac{(1 - v_i{}^2)}{E_i} \tag{7}$$

where v is the Poisson's ratio, subscript s and i refer to the sample and indenter material respectively.

The above approach works very well for typical engineering materials and some hard biomaterials (e.g., bone [59]), but for the materials with time-independent mechanical responses (i.e., soft biomaterials here), Oliver–Pharr method is invalidated due to the creep to overwhelm the elastic recovery [60], resulting in a near-vertical or even negative slope in the initial unloading region. For this reason, some "corrections" have been adopted based on Oliver–Pharr analysis, such as high unloading rates [61], long hold periods [62] and the data analysis based on the measured creep rate [63–65]. Another limitation of Oliver–Pharr method is that the results necessarily rely on the contact area (tip area function), which may lead to significant errors in the cases of the "tip radius" effect [66,67] and the pile-up (or sink-in) effect [68]. Accordingly, some correction factors [69,70] or new approaches [71] may be also taken into account to evaluate the actual properties of the material.

3.3.2. Hertz Contact Model

Hertz model, the most well-known and applied theory in mechanical characterization of materials, was proposed by Hertz to solve the problem of contact between two smooth, ellipsoidal bodies [72]. Some assumptions employed for validity are required:

- Small contact area and small deformations.
- Isotropic and homogenous materials.
- Adhesionless and frictionless surfaces.

Following the above assumptions, Sneddon made a significant contribution to the theoretical framework to formulate the relationship between force and depth for a punch of arbitrary profile penetrated [73]. Initially, Sneddon's solution was developed for elastic contact of hard materials, and it has been proved applicable to extend to determine the initial shear modulus and even hyperelastic parameters of non-linear soft materials [74].

For the indentation by a flat-ended cylindrical indenter, the indentation force P and depth h can be directly related through the following equation:

$$P_{\text{cylinder}} = \frac{4R}{1 - v} Gh \tag{8}$$

where R is the cylinder radius, and G and v are the shear modulus and Poisson's ratio of the soft biomaterial, respectively. For a spherical indenter (Figure 5), the indentation force is:

$$P_{\text{sphere}} = \frac{8\sqrt{R}}{3(1-v)} Gh^{3/2} \tag{9}$$

where R is the sphere radius, and for a conical indenter:

$$P_{\text{cone}} = \frac{4\tan\alpha}{\pi(1-v)} Gh^2 \tag{10}$$

where α is the cone half angle.

Sometimes, the above equations may need to be modified owing to the non-ideal effects while mechanical characterization [31,75], such as the correction factor of size effect [76,77] and the compensating factor for the tip [78,79]. When it comes to micro/nano-scale (such as cell indentation), the above effect may be much more significant considering that the indentation depth is comparable to the cell dimension. The Hertz contact model may become invalid owing to large deformation or thin-layer effect, in which case some correction factors or new models need to be taken into account [80–82].

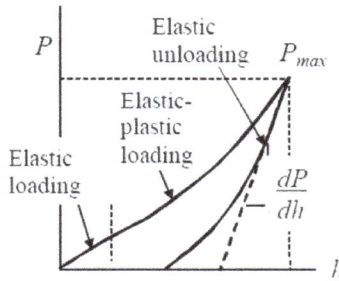

Figure 5. Typical force–indentation ($P - h$) curve for spherical indenter, in which the Oliver–Pharr model can be used in elastic unloading portion, and Hertz contact model can be used in elastic loading portion. Reproduced and adapted with permission from [49].

For the characterization of time-dependent behavior, a Boltzmann hereditary integral, based on associated solution for a linearly elastic material, was proposed to capture the time-dependent stresses and deformations along the total temporal scale [83]. For the flat-ended cylindrical indenter with a radius R, applying the hereditary integral in Equation (8), the time-dependent indentation force $P(t)$ and depth h relation is represented by:

$$P_{\text{cylinder}}(t) = \frac{4R}{1-v} \int_0^t G(t-\tau)\left(\frac{dh}{d\tau}\right) d\tau \tag{11}$$

Similarly, for the spherical indenter with a radius R:

$$P_{\text{sphere}}(t) = \frac{8\sqrt{R}}{3(1-v)} \int_0^t G(t-\tau)\left(\frac{dh^{3/2}}{d\tau}\right) d\tau \tag{12}$$

and for the conical indenter with a cone half angle α:

$$P_{\text{cone}}(t) = \frac{4\tan\alpha}{\pi(1-v)} \int_0^t G(t-\tau)\left(\frac{dh^2}{d\tau}\right) d\tau \tag{13}$$

In Equations (11)–(13), the time-dependent shear modulus $G(t)$ can be expressed by a linear viscoelastic model (see Equations (1)) to characterize time-dependent mechanical properties [84,85].

Overall, compared with the Oliver–Pharr model based on elastic–plastic deformation (or the "correction" based on viscous-elastic–plastic deformation), the Hertz and Sneddon models, based on elastic contact, may dominate the literature for soft biomaterials due to their simplicity and widespread application [86]. The comparisons between the Oliver–Pharr method and Hertz contact model are summarized in Table 1, in which some aspects, such as tip selection and control mode, will be further discussed in the next section. The reader is also encouraged to review more analytical approaches in/beyond the Hertzian regime provided in the literature [87].

Table 1. Comparisons between the Oliver–Pharr method and Hertz contact model.

	Oliver–Pharr Model	Hertz Contact Model
Application	Typical engineering materials and hard biomaterials	Soft biomaterials
Tip Selection	Typically using sharp indenters (e.g., Berkovich and Vickers)	Typically using dull indenters (e.g., spherical and flat-ended)
Control Mode	Typically using load-control	Typically using displacement-control
Data Analyzed	Unloading	Loading (include holding when characterizing viscoelasticity)
Method for Time-Dependence	"Correfcted" Oliver–Pharr analysis	Boltzmann hereditary integral (Equations (11)–(13))

4. Test Protocols

Compared with the typical engineering materials and hard biomaterials, soft biomaterials are much more compliant, with the elastic modulus values down to MPa or even kPa range. Considering the inherent complexities of soft biomaterials, great care and consideration are required to characterize mechanical behavior at multiple scales, especially when utilizing commercial instrumentation. In this section, some limitations as to data collection and analysis are considered, and a set of guidelines are presented.

4.1. Tip Selection

4.1.1. Tip Material

As is well-known, the materials of indenter tip are relatively stiff compared with the sample, to ensure that the compliance of tip is tiny and can be ignored. Owing to the low modulus of soft biomaterials, tip materials can be changed from the diamond or sapphire, which is widely used in typical nanoindentaion, to steel [88], aluminum [33], or even silica [89] and silicon [31].

4.1.2. Tip Geometry

For indenting soft biomaterials, whose force–indentation regime is opposite to stiff materials, small forces arise at relatively large displacements compared with large forces at relatively small displacements [90]. Thus, a dull tip (spherical [43] and flat-ended [33]) is commonly used instead of a sharp tip (Berkovich [91], Vickers [92], conical [93], and cube corner [94]). Some advantages of a dull tip are as follows:

- A dull tip can achieve a larger contact area and therefore a higher force level compared with a sharp indenter at the same indentation depth, which allows for the testing within range/resolution of the instrument.

- The high plastic deformations and stress concentrations induced by a sharp tip could lead to sample damage (from tissue penetration to cell membrane rupture) and excess of linear elastic limit.
- Considering heterogeneity and complexity of soft biomaterials inherently, a sharp tip may lead to significant differences among repetitive tests owing to too small tip-sample contact area. Thus, one measurement with a dull tip can be considered as the average of many measurements with the sharp probe, and consequently, less time to obtain robust statistics.

In spite of the wide application of spherical and flat-ended tips for characterizing soft biomaterials, each one has its own advantages and drawbacks, and may be more applicable to specific conditions, as shown in Table 2. The reader is also encouraged to review more details about different tips in the literature [82,95].

Table 2. Comparison between sphere and flat-ended tip for characterization of soft biomaterials

	Sphere Tip	Flat-Ended Tip
Advantages	• Can offer the stress without high stress concentration at the contact perimeter.	• Can simplify data analysis within a constant contact area during indentation. • Can achieve more force value compared with sphere tip at the same depth, which is crucial to the instruments with low signal-to-noise ratios.

4.1.3. Tip Dimension

By virtue of different tip dimensions, indentation instruments can provide the versatility of functioning across the length spectrum ranging from bulk scale to the micro/nano-scale, and even large-scale indentation (with millimeter-sized or larger tips) has also been widely used for characterization of some extremely soft biomaterials (e.g., brain tissue [33]). Anyway, to avoid complicating the interpretation of the data, a criterion is suggested here to be exercised while indentation: the size of tip $R \ll$ current length scale level of the measured sample $L_{current\ level}$, and tip size $R \gg$ the lower length scale level $L_{lower\ level}$. For example, if the goal is to characterize the mechanical behavior at tissue-level, the tip size need to be much small compared with the tissue, to adapt to the Hertz assumption of infinite half space ($R \ll L_{tissue\ level}$), as well as larger than the diameter of an individual cell or fiber ($R \gg L_{cell\ level}$).

4.2. Control Mode

For most commercial nanoindenters, a load-controlled mode is utilized by default. However, for soft biomaterials which are extremely compliant, adhesive and time-dependent, this mode can be difficulties with the contact detection and data analysis, in which case displacement-control is extremely useful.

Firstly, displacement-control can overcome unambiguous tip-sample contact detection. Whether the contact is ascertained by a small force change or a small apparent stiffness change in the mode of load-control, the characteristics of low modulus and time-dependence of soft biomaterials would play against the change of force and stiffness respectively. Accordingly, tests starting below the sample surface detection may happen, leading to significant overestimations of the elastic modulus of soft materials via indentation [96]. On the contrary, the test can be initiated prior to tip–sample contact in the displacement-controlled mode, when the probe is slightly above the sample, and some correction of the contact point can be applied post-hoc.

Additionally, displacement-control can simplify data analysis of time-dependent behavior. For most soft biomaterials, they are viscoelastic materials whose properties depend on the strain rate. In the mode of load-control, the strain rate is not a constant during indentation due to displacement creeping, which complicates the mechanical characterization of soft biomaterials.

4.3. Transition from Linear to Nonlinear Behavior

In general, the assumption of small strain need to be satisfied when characterizing soft biomaterials based on Hertz contact model, so it is not trivial to establish criteria for linear behavior threshold. Albeit some plausible criterions have been proposed empirically, such as a threshold of 10% indentation strain [97], it remains vague and unclear for the transition from linear to nonlinear stress–strain behavior [15]. Actually, the ranges of linearity differ from different materials and even testing methods. For example, some studies limit the strain values below 1% to guarantee the linear behavior for brain tissue via dynamic shearing testing [48,98]. As for indentation of brain tissue, this level is increased up to 10% strain by Elkin et al. [99,100], or even 45% strain adopted by MacManus et al. [31]. More recently, Budday et al. concluded different linear ranges of brain tissue in one study, in which the threshold was limited to 10% strain in compression or tension, and 20% in shear [101].

The situations are equally complicated when it comes to microscale. Leipzig and Athanasiou characterize elastic behavior of chondrocytes at ~30% strain via compression [102], and a lower level (15% strain) is suggested to be applicable in AFM indentation according to Darling et al. [103]. More strictly, this value is limited to 5% by Chen and Lu [80].

Hence, some criteria of linear behavior threshold should be referred cautiously, especially for some high levels, and be explored specifically depending on different materials and testing methods. Additionally, some correction factors or new models, extending beyond the Hertzian and linear elastic models, can be developed and utilized to minimize the above issues [80].

4.4. Adhesion and Point of Contact (POC)

Adhesion, the most prevalent form of tip–sample interaction in the indentation of soft biomaterials, may omit the true point of contact erroneously, which in turn interferes the measurements of mechanical characterization (Figure 6). It is suggested that the adhesion between the tip and the sample is a significantly important parameter and needs to be taken into account for mechanical characterization, whether for a dedicated nanoindenter or AFM indentation [104,105].

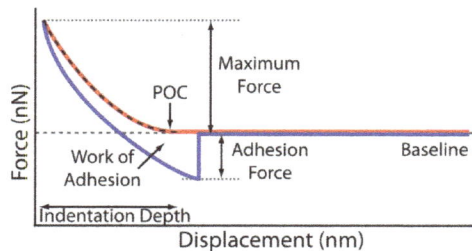

Figure 6. Schematic diagram of the adhesion in AFM nanoindentation, in which the red line is the approach and the blue line is the retract curve. Reproduced with permission from [86].

When adhesion is present, some non-interactive contact models, such as the aforementioned Oliver–Pharr and Hertz contact models, may need to be modified to make the modulus values more accurate [106]. This was pioneered by JKR (Johnson–Kendall–Roberts) theory [107], which introduces an apparent Hertz load or the equivalent load. Subsequently, a seemingly contradictory theory, DMT (Derjaguin–Muller–Toporov) theory [108], was proposed, which is assumed to follow the Hertz model. Actually, The above two theories were identified in terms of sample compliance adhesive force range, in which JKR theory describes the case of relatively compliant materials with large contact size and adhesive force, and DMT theory stands the opposite with stiff materials, small contact size, and adhesive force [109]. Further, the adhesive contact mechanics was developed by Maugis–Dugdale (MD) theory, which is an intermediate case spanning the JKR and DMT limits [110,111].

Aside from the theories based on adhesive contact, some methods for non-adhesive contact have been employed for POC determination, and summed up in the literature [86,87]. These include, but are not limited to, visual inspection [112], model fitting [105], extrapolation [112,113], and Bayesian analysis [114]. Additionally, another strategy has been considered to characterize mechanical behavior without needing to determine the POC. A method was initially proposed by A-Hassan et al. to compare a known reference sample [115]. Later a protocol of data processing to linearize the Hertz and Sneddon equations for different tips has been developed and used widely [33,116]. According to Equations (8)–(10), shear modulus G can be determined by measuring the slope of indentation force for differen tips $P_{cylinder}$, $(P_{sphere})^{2/3}$ and $(P_{cone})^{1/2}$ versus indentation depth h:

$$G_{cylinder} = \frac{1}{4}\text{slope}\frac{1-\nu}{R} \tag{14}$$

$$G_{sphere} = \frac{3}{8}\text{slope}^{3/2}\frac{1-\nu}{\sqrt{R}} \tag{15}$$

$$G_{cone} = \frac{\pi}{4}\text{slope}^2\frac{1-\nu}{\tan\alpha} \tag{16}$$

In this way, the shear modulus, defined in terms of remaining constant parameters of tip geometric and Poisson's ratio, can be calculated without POC.

4.5. Viscoelasticity and Poroelasticity

It is known that almost all soft biomaterials exhibit time-dependent behavior. Apart from some aforementioned typical indentation methods to characterize time-dependent behavior, dynamic testing, where a sinusoidal load rather than a trapezoidal load is applied to measure storage and loss modulus directly as a function of loading frequency, is also widely used [117,118].

Consider that many soft biomaterials exhibit time-dependent behavior under specific physiological conditions, some explicit analysis connected with physiological conditions are more likely to be central to future studies of soft biomaterials, rather than considering water solely as a solvent or an adaptive component. Compared with linear viscoelasticity, poroelasticity can clearly model the multiphase nature of soft biomaterials. Additionally, different from the empirical fitting of linear viscoelastic model, poroelastic model is mechanistic and can relate the rheological properties to structural or biological parameters, which in turn can predict the changes in rheology due to microstructural changes [32]. It also need to be mentioned that these two effects (viscoelasticity and poroelasticity) act relatively independently and can be separated uniquely with exhibiting both behaviors simultaneously [119]. This independence makes sense since one mechanism is dictated by tissue dissipation and the other by fluid flow [26].

In the case of some complicated constitutive response, such as a nonlinear viscoelastic or a poroelastic model, there may be no closed-form analytical expression of the indentation issue, and some approaches of finite element analysis (FEA) are proved to be useful [90]: inverse optimization FEA model [120], approximation of forward FEA simulations [121], and a fitted database [122]. In fact, no matter what constitutive model is used to characterize time-dependent behavior, most data analysis are performed "off-line" from the instrument software, which impedes development of commercial instruments, especially in the field of biomaterials characterization. More details about the comparisons between viscoelastic model and poroelastic model can be seen in [82].

4.6. Sample Hydration and Environmental Control

Another critical factor for mechanical characterization of most soft biomaterials is the hydration state of the sample. Owing to the difficulties respect to instrument setup, data acquisition and analysis in handling the hydration state of the sample, there have been to date, relatively few studies under a hydrated state. However, a number of studies have concluded lower elastic modulus values for hydrated soft biomaterials with as much as an order of magnitude than that of the dried counterpart [123,124].

There are two different basic strategies for hydrated testing of samples: "off-line" and "on-line". Samples can be hydrated in fluid prior to testing, and just removed from the fluid while indentation within a short time period before the fluid evaporates [88,118]. This is the simplest hydration method, but may be problematic in a way considering the susceptibility of biomaterials to ambient conversion. Alternatively, samples can also be tested in a hydrated or submerged fluid environment, including partially/fully submerged in fluid [33,42], surrounded by hydrating foam layer [29], hydrated by specialized irrigation system [51], and covered by physiological saline-coated gauze [30].

Even if the samples have been hydrated, it is still unmet for environmentally-controlled nanoindentation under the truly physiological state. First is the types of fluid environments. Sometimes, water is used for sample hydration instead of physiological saline or other special solution (such as artificial cerebrospinal fluid for brain tissue) to minimize the risk of fluid damage to the instrument, especially for commercial instruments, but such substitution may lead to alteration of mechanical response [125]. Further, most studies are performed at room temperature rather than body temperature, and even current models for mechanical characterization perform poorly when fitting such experimental data. Thus, some control of experimental design and better models are needed to characterize soft biomaterials under a truly physiological state.

Consequently, considering the complexity of physiological activity, it may be better to some extent to achieve environmentally controlled nanoindentation based on "biology" instead of "mechanics", i.e., it is more intriguing for in vivo testing, compared with in situ or in vitro testing.

5. Applications

5.1. AFM Indentation

By virtue of its high spatial and force resolutions, AFM has long been a critical technique for characterizing biological systems, from biomolecules to complexes, which, in turn, has been a driving force for the expansion of the repertoire of novel applications [126–128]. Today, AFM has become increasingly sophisticated, especially combined with surface morphology and topography at the micro/nano-scale [128–130].

5.1.1. Biomolecular Level

Mechanical characterization at the biomolecular level can provide the insight into how properties and functions at more intricate biomolecular complexes (single cells and tissues, and organelles) arise from a myriad of single biomolecules. For example, intermediate filaments offer mechanical stability to cells when exposed to mechanical stress, and act as a support when the other cytoskeletal filaments cannot keep the structural integrity of the cells [131]. AFM can also be applied for the mapping of target RNA distribution using probe DNA-immobilized AFM tips [132], which is fundamentally important to understand the regulatory mechanisms underlying cellular and tissue differentiation [133].

In addition to the single biomolecules, interactions of molecules (Figure 7) are also able to produce significant effects on the functionality of larger-scale biomolecular complexes and tissues. DNA-protein interactions, which play important roles in numerous fundamental biological processes such as DNA replication, packing, recombination, DNA repair, RNA transport, and translation, can also be explored by AFM [134].

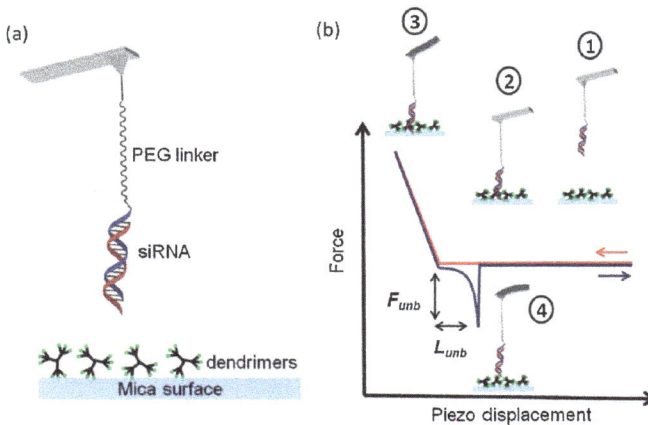

Figure 7. (a) Scheme of the tip functionalization and dendrimer adsorption on mica. (b) Main steps of a force curve depicting a molecular recognition (specific) event. ①. Tip far from the surface. ②. Initial tip–surface contact (approaching). ③. Tip–surface repulsive region. ④. Molecular recognition unbinding force. Reproduced with permission from [135].

Further, characterization of biomolecules can also provide the potential to supplement the field of biomaterials with new resources for the design of novel materials and nanocomposites [136].

5.1.2. Cellular Level

Cell mechanics, one of the most important biophysical properties, can provide new perspectives on pathologies and classic biological research questions [32]. AFM has been recognized as a powerful technique to understand cellular processes in the applications of regenerative medicine and tissue engineering [137]. Further, AFM indentation can offer various functionalities, starting from surface imaging to detection of interaction forces, delivering quantitative mechanical behavior that can describe changes characteristic in diagnosis of many pathological conditions, such as cancer [138]. Together with surface topography, AFM can also offer accurate constitutive data about surface physico-chemical properties, such as elasticity [139], friction [140], adhesion [141], and even nanoscale surgeries on live cells (Figure 8) [20].

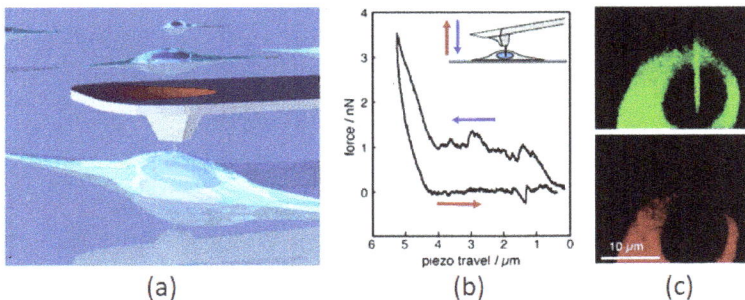

Figure 8. Nanoscale operation of a living cell using AFM. (a) Schematic representation of a nanoneedle on the AFM tip over a living cell. (b) Force-distance curves during approach and retraction from a melanocyte. (c) Cross-section images for green and red emission processed from confocal slices. Reproduced and adapted with permission from [20].

In addition to the cell itself, the extracellular matrix (ECM) and the cell membrane involved in communications between cells and surrounding matrix also has a critical role in regulating cell properties and behavior. AFM can provide important insights on our current understanding of the mechanics of cells, ECM, and cell-ECM bidirectional interactions [142–145].

The reader interested in more details of specific cells can refer to a number of AFM indentation publications including: cardiomyocytes [146], liver cells [147], chondrocytes [148], neurons [149], fibroblasts [150], and mesenchymal stem cells [151].

5.2. Dedicated Instrumented Indentation

Owing to the indirect loading and the compliance of cantilever combined with adhesive effects of AFM indentation [90], more robust data may be provided via dedicated instrumented indentation, especially at the tissue level (Figure 9).

For example, in the field of traumatic brain injury (TBI), detailed anatomically correct geometry and accurate constitutive data of brain tissue are required to construct accurate brain models [31]. Brain models of TBI play an important role in the process of evaluating and understanding the complex physiologic, behavioral, and histopathologic changes associated with TBI, which, in turn, is necessary to establish numerical predictions and new therapeutic strategies in brain injury [152].

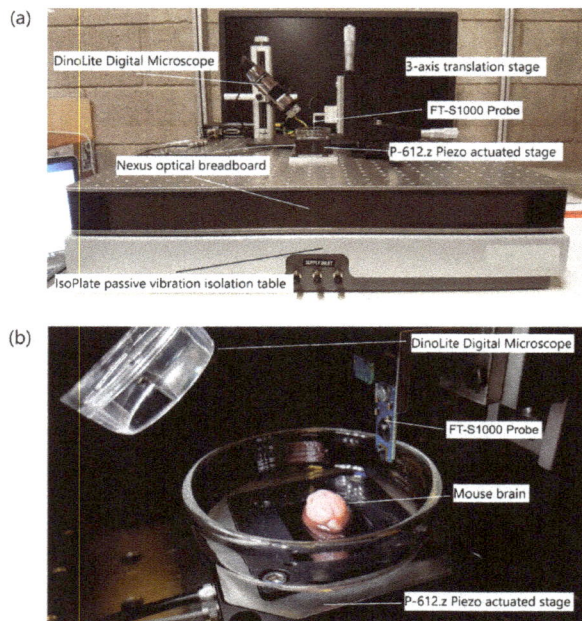

Figure 9. (a) Major components of the custom built micro-indentation experimental apparatus, and (b) a close-up of the force sensing probe and mouse brain specimen prior to indentation. Reproduced with permission from [153].

Indentation technique can also be utilized in the fields of tissue replacement and repair. It is shown that nanoindentation can be utilized for functional evaluation of cartilage repair tissue [30]. Owing to the in situ testing and nondestructive nature, indentation technique allows for subsequent histological or biochemical analysis of the same tissue, which offers a more complete description of repair tissue.

Other applications involved in indentation to soft tissue include, but are not limited to, poroelastic properties of articular cartilage [154], viscoelastic properties of liver tissues [155], remineralization of demineralized dentin [156], and in vivo tests of skin [157].

6. Challenges and Perspectives

6.1. Open Questions and Challenges

Nanoindentation techniques have emerged as indispensable tools for mechanical characterization, with the development of the versatility. However, apart from some inherent characteristics we have mentioned above, there are still a number of challenges which are particularly significant in soft biomaterials.

Firstly, and the most importantly, there is a paucity of consensus on the appropriate data analysis to mechanical characterization of soft biomaterials. This issue can be illustrated by twofold factors:

- Deviations among individual measurements in one research, owing to diversities in tissue architecture and structural heterogeneity (Figure 10).
- Very large disparities in the results reported in different studies, essentially linked with variations in the protocols [158].

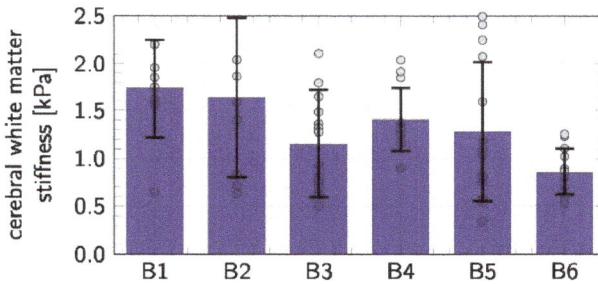

Figure 10. Example of deviations among individual measurements in the literature. $n = 116$ indentation tests of cerebral white matter within six brains (B1–B6). Individual measurements are indicated as dots; means and standard deviations are shown as bar graphs. Reproduced with permission from [17].

As such, it remains difficult to characterize and model soft biomaterials accurately, even though there is a very large amount of information regarding mechanical characterization. The influence of the first factor which is attributed to inherent complexities of soft biomaterials, can be reduced to a certain extent by some repetitive experiments. In this regard, it seems to be more urgent to build a standardized routine from sample preparation to test execution, hence, properly dealing with the large disparities reported in different studies.

Secondly, the surface of some soft biomaterial is not flat and smooth ideally that comes in the form of numerous asperities (Figure 11). It may invalidate the underlying assumption of some of analytical theories, as well as bring some difficulties to test operation and data analysis, especially for a dull indenter [72]. In this case, a relative large indentation depth may be applicable to some extent [159,160]. It need to be mentioned that although some additional processing and preparation methods (e.g., cryomicrotoming) can be adopted to reduce surface roughness, the mechanical properties of the sample may be affected by the preparation or processing mechanism. Further, it seems to be more fascinating for biomaterials to interpret the mechanical behavior thereof with original morphology, i.e., in situ and in vivo tests.

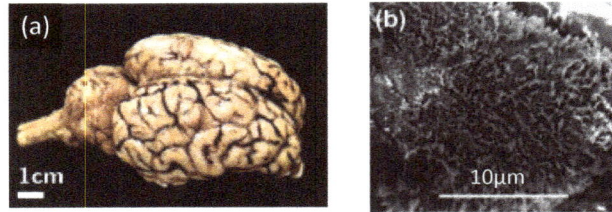

Figure 11. Examples of the non-ideal surface of soft biomaterials in the form of numerous asperities. (**a**) Macroscale: the surface of porcine brain tissue with sulci and gyri, reproduced and adapted with permission from [161]; and (**b**) microscale: the surface of human cervical epithelial cells with brush, reproduced and adapted with permission from [162].

Thirdly, some assumptions of analytical relationships, such as isotropy and homogeneity, are poor descriptions and are never perfectly met in practice. The application of above analytical models may lead to certain degree of error herein. Since these aspects are directly related to the microstructural features and variations thereof, an analytical model that could simultaneously account for all involved complexities of soft biomaterials, such as nonlinear behavior, time-dependence, anisotropy, and heterogeneity, may be far-reaching, but also be daunting due to the very large workload and specialized knowledge.

6.2. Perspectives and Opportunities

Despite significant progress based on current research-driven user base, there is a general agreement that a more detailed understanding of mechanics involved in accurate constitutive data and anatomical distribution is required, thereby offering more opportunities for medical field such as pathology and biochemistry.

For many soft biomaterials, mechanical characterization under in vivo environment may have more important consequences in understanding the physical biology. Compared with the test ex vivo, some alterations in mechanical properties may exist due to perfusion pressure, tissue degradation, and boundary condition effects [158]. More recent study indicates that the mechanical properties change drastically within only several minutes post mortem [163]. Accordingly, in vivo testing may be a better (even the only) strategy to provide insight into mechanical behavior of some soft biomaterials.

The investigation of the testing under real service conditions and even some extreme environment conditions is another critical step for future research. In fact, most soft biomaterials operate at multiple and complicated loading conditions, and thus some measurements under ideal experiment condition may overlook many potential details. In particular, some results derived under one loading-boundary mode does not necessarily conform to the material response under another mode [161]. In addition, the biomaterials under some extreme environment conditions, such as high electric field, high magnetic fields, intense radiation, and extreme low/high temperature, may have some novel behavior [164], which, in turn, provide a better understanding for the application in some fields, such as medical research and space exploration.

Compared with most researches focused on the mechanical characterization, naoindentation can also be developed as a powerful tool in the field of mechanobiology [165]. Recent findings have directly linked development, cell differentiation, physiology, and disease to mechanical response both at the cellular and tissue levels [166]. Insights into the mechanical signal transmission from the macroscale to the nanoscale would allow us to better understand composition–structure–properties relationships in biomaterials.

Elastographic techniques (e.g., magnetic resonance and ultrasound elastography), also acting as noninvasive tools, can quantify the mechanical behavior of soft biomaterials in vivo as well [163].

Despite the infancy of these techniques, lacking a thorough verification and validation [167], it is expected that broad application prospects will arise in the fields of various pathologies if connected with nanoindentation techniques.

7. Conclusions

In this review, we have discussed the topic of nanoindentation of soft biological materials. Owing to some characteristics of soft biomaterials, including nonlinearity, time-dependence, anisotropy, and heterogeneity, some test protocols regarding constitutive model and data analysis are different from typical engineering materials. Although a standardized routine for nanoindentation of biomaterials remains to be built, nanoindentation techniques have emerged as indispensable tools for investigation of both biomechanics and mechanobiology. The applications of nanoindentation have contributed far more than simply characterizing mechanical behavior from bulk scale to the micro/nano-scale. There will be more opportunities for their use in other fields, such as physiology and pathologies.

Author Contributions: H.Z. conceived and developed the ideal of this work; L.Q. drifted the paper and revised the manuscript; all authors have given approval to the final version of the manuscript.

Funding: This research was funded by the National Key R and D Program of China (2018YFF01012400), the National Science and Technology Innovation Leading Academic (Ten Thousand Talent Program), the Fund Guiding on Strategic Adjustment of Jilin Provincial Economic Structure Project (2014Z045), Major project of Jilin Province Science and Technology development plan (20150203014GX), the Special Fund Project of Jilin Provincial Industrial Innovation (2016C030), the Jilin Provincial Middle and Young Scientific and Technological Innovation Talent and Team Project (20170519001JH), and the Program for JLU Science and Technology Innovative Research Team (2017TD—04).

Conflicts of Interest: The authors declare no conflict of interest.

References

1. Zadpoor, A.A. Mechanics of Biological Tissues and Biomaterials: Current Trends. *Materials* **2015**, *8*, 4505–4511. [CrossRef] [PubMed]
2. Garcia-Manyes, S.; Domenech, O.; Sanz, F.; Montero, M.T.; Hernandez-Borrell, J. Atomic force microscopy and force spectroscopy study of Langmuir-Blodgett films formed by heteroacid phospholipids of biological interest. *Biochim. Biophys. Acta Biomembr.* **2007**, *1768*, 1190–1198. [CrossRef] [PubMed]
3. Huang, S.; Ingber, D.E. Cell tension, matrix mechanics, and cancer development. *Cancer Cell* **2005**, *8*, 175–176. [CrossRef] [PubMed]
4. Discher, D.E.; Janmey, P.; Wang, Y.L. Tissue cells feel and respond to the stiffness of their substrate. *Science* **2005**, *310*, 1139–1143. [CrossRef] [PubMed]
5. Engler, A.J.; Sen, S.; Sweeney, H.L.; Discher, D.E. Matrix elasticity directs stem cell lineage specification. *Cell* **2006**, *126*, 677–689. [CrossRef] [PubMed]
6. Stolz, M.; Gottardi, R.; Raiteri, R.; Miot, S.; Martin, I.; Imer, R.; Staufer, U.; Raducanu, A.; Duggelin, M.; Baschong, W.; et al. Early detection of aging cartilage and osteoarthritis in mice and patient samples using atomic force microscopy. *Nat. Nanotechnol.* **2009**, *4*, 186–192. [CrossRef] [PubMed]
7. Cloots, R.J.H.; van Dommelen, J.A.W.; Kleiven, S.; Geers, M.G.D. Multi-scale mechanics of traumatic brain injury: Predicting axonal strains from head loads. *Biomech. Model. Mechanobiol.* **2013**, *12*, 137–150. [CrossRef] [PubMed]
8. Oyen, M.L. The materials science of bone: Lessons from nature for biomimetic materials synthesis. *MRS Bull.* **2008**, *33*, 49–55. [CrossRef]
9. Rashid, B.; Destrade, M.; Gilchrist, M.D. Mechanical characterization of brain tissue in tension at dynamic strain rates. *J. Mech. Behav. Biomed.* **2014**, *33*, 43–54. [CrossRef]
10. Buckley, M.R.; Gleghorn, J.P.; Bonassar, L.J.; Cohen, I. Mapping the depth dependence of shear properties in articular cartilage. *J. Biomech.* **2008**, *41*, 2430–2437. [CrossRef]
11. Li, Y.H.; Hong, Y.; Xu, G.K.; Liu, S.B.; Shi, Q.; Tang, D.D.; Yang, H.; Genin, G.M.; Lu, T.J.; Xu, F. Non-contact tensile viscoelastic characterization of microscale biological materials. *Acta Mech. Sin.* **2018**, *34*, 589–599. [CrossRef]

12. MacManus, D.B.; Gilchrist, M.D.; Murphy, J.G. An empirical measure of nonlinear strain for soft tissue indentation. *R. Soc. Open Sci.* **2017**, *4*, 170894. [CrossRef] [PubMed]
13. Cuvelier, D.; Derenyi, I.; Bassereau, P.; Nassoy, P. Coalescence of membrane tethers: Experiments, theory, and applications. *Biophys. J.* **2005**, *88*, 2714–2726. [CrossRef] [PubMed]
14. Hayashi, K.; Iwata, M. Stiffness of cancer cells measured with an AFM indentation method. *J. Mech. Behav. Biomed.* **2015**, *49*, 105–111. [CrossRef] [PubMed]
15. Kurland, N.E.; Drira, Z.; Yadavalli, V.K. Measurement of nanomechanical properties of biomolecules using atomic force microscopy. *Micron* **2012**, *43*, 116–128. [CrossRef] [PubMed]
16. Bouchonville, N.; Meyer, M.; Gaude, C.; Gay, E.; Ratel, D.; Nicolas, A. AFM mapping of the elastic properties of brain tissue reveals kPa mu m(-1) gradients of rigidity. *Soft Matter* **2016**, *12*, 6232–6239. [CrossRef] [PubMed]
17. Weickenmeier, J.; de Rooij, R.; Budday, S.; Steinmann, P.; Ovaert, T.C.; Kuhl, E. Brain stiffness increases with myelin content. *Acta Biomater.* **2016**, *42*, 265–272. [CrossRef]
18. Stewart, M.P.; Helenius, J.; Toyoda, Y.; Ramanathan, S.P.; Muller, D.J.; Hyman, A.A. Hydrostatic pressure and the actomyosin cortex drive mitotic cell rounding. *Nature* **2011**, *469*, 226–230. [CrossRef]
19. Prevost, T.P.; Jin, G.; de Moya, M.A.; Alam, H.B.; Suresh, S.; Socrate, S. Dynamic mechanical response of brain tissue in indentation in vivo, in situ and in vitro. *Acta Biomater.* **2011**, *7*, 4090–4101. [CrossRef]
20. Obataya, I.; Nakamura, C.; Han, S.; Nakamura, N.; Miyake, J. Nanoscale operation of a living cell using an atomic force microscope with a nanoneedle. *Nano Lett.* **2005**, *5*, 27–30. [CrossRef]
21. Chaudhuri, O.; Parekh, S.H.; Lam, W.A.; Fletcher, D.A. Combined atomic force microscopy and side-view optical imaging for mechanical studies of cells. *Nat. Methods* **2009**, *6*, 383. [CrossRef] [PubMed]
22. Friedrichs, J.; Legate, K.R.; Schubert, R.; Bharadwaj, M.; Werner, C.; Mullner, D.J.; Benoit, M. A practical guide to quantify cell adhesion using single-cell force spectroscopy. *Methods* **2013**, *60*, 169–178. [CrossRef] [PubMed]
23. Lin, D.C.; Shreiber, D.I.; Dimitriadis, E.K.; Horkay, F. Spherical indentation of soft matter beyond the Hertzian regime: Numerical and experimental validation of hyperelastic models. *Biomech. Model. Mechanobiol.* **2009**, *8*, 345–358. [CrossRef] [PubMed]
24. Olesiak, S.E.; Sponheimer, M.; Eberle, J.J.; Oyen, M.L.; Ferguson, V.L. Nanomechanical properties of modern and fossil bone. *Palaeogeogr. Palaeocl.* **2010**, *289*, 25–32. [CrossRef]
25. Creton, C.; Ciccotti, M. Fracture and adhesion of soft materials: A review. *Rep. Prog. Phys.* **2016**, *79*, 046601. [CrossRef] [PubMed]
26. Oyen, M.L. Nanoindentation of hydrated materials and tissues. *Curr. Opin. Solid State Mater. Sci.* **2015**, *19*, 317–323. [CrossRef]
27. Parra, A.; Casero, E.; Lorenzo, E.; Pariente, F.; Vazquez, L. Nanomechanical properties of globular proteins: Lactate oxidase. *Langmuir* **2007**, *23*, 2747–2754. [CrossRef]
28. Li, Q.S.; Lee, G.Y.H.; Ong, C.N.; Lim, C.T. AFM indentation study of breast cancer cells. *Biochem. Biophys. Res. Commun.* **2008**, *374*, 609–613. [CrossRef]
29. Ebenstein, D.M.; Pruitt, L.A. Nanoindentation of soft hydrated materials for application to vascular tissues. *J. Biomed. Mater. Res. A* **2004**, *69*, 222–232. [CrossRef]
30. Ebenstein, D.M.; Kuo, A.; Rodrigo, J.J.; Reddi, A.H.; Ries, M.; Pruitt, L. A nanoindentation technique for functional evaluation of cartilage repair tissue. *J. Mater. Res.* **2004**, *19*, 273–281. [CrossRef]
31. MacManus, D.B.; Pierrat, B.; Murphy, J.G.; Gilchrist, M.D. Dynamic mechanical properties of murine brain tissue using micro-indentation. *J. Biomech.* **2015**, *48*, 3213–3218. [CrossRef] [PubMed]
32. Moeendarbary, E.; Harris, A.R. Cell mechanics: Principles, practices, and prospects. *Wiley Interdiscip. Rev. Syst. Biol. Med.* **2014**, *6*, 371–388. [CrossRef] [PubMed]
33. Qian, L.; Zhao, H.W.; Guo, Y.; Li, Y.S.; Zhou, M.X.; Yang, L.G.; Wang, Z.W.; Sun, Y.F. Influence of strain rate on indentation response of porcine brain. *J. Mech. Behav. Biomed.* **2018**, *82*, 210–217. [CrossRef] [PubMed]
34. Falland-Cheung, L.; Scholze, M.; Hammer, N.; Waddell, J.N.; Tong, D.C.; Brunton, P.A. Elastic behavior of brain simulants in comparison to porcine brain at different loading velocities. *J. Mech. Behav. Biomed.* **2018**, *77*, 609–615. [CrossRef] [PubMed]
35. Jia, F.; Ben Amar, M.; Billoud, B.; Charrier, B. Morphoelasticity in the development of brown alga Ectocarpus siliculosus: From cell rounding to branching. *J. R. Soc. Interface* **2017**, *14*, 20160596. [CrossRef] [PubMed]

36. MacManus, D.B.; Pierrat, B.; Murphy, J.G.; Gilchrist, M.D. Mechanical characterization of the P56 mouse brain under large-deformation dynamic indentation. *Sci. Rep.* **2016**, *6*, 21569. [CrossRef] [PubMed]
37. Marckmann, G.; Verron, E. Comparison of hyperelastic models for rubber-like materials. *Rubber Chem. Technol.* **2006**, *79*, 835–858. [CrossRef]
38. Pan, Y.H.; Zhan, Y.X.; Ji, H.Y.; Niu, X.R.; Zhong, Z. Can hyperelastic material parameters be uniquely determined from indentation experiments? *RSC Adv.* **2016**, *6*, 81958–81964. [CrossRef]
39. Crichton, M.L.; Chen, X.F.; Huang, H.; Kendall, M.A.F. Elastic modulus and viscoelastic properties of full thickness skin characterised at micro scales. *Biomaterials* **2013**, *34*, 2087–2097. [CrossRef] [PubMed]
40. Oyen, M.L. Indentation of nonlinearly viscoelastic solids. *Fundam. Nanoindent. Nanotribol.* **2008**, *1049*, 99–104. [CrossRef]
41. Wang, H.F. *Theory of Linear Poroelasticity with Applications to Geomechanics and Hydrogeology*; Princeton University Press: Princeton, NJ, USA, 2017.
42. Oyen, M.L.; Shean, T.A.V.; Strange, D.G.T.; Galli, M. Size effects in indentation of hydrated biological tissues. *J. Mater. Res.* **2012**, *27*, 245–255. [CrossRef]
43. Li, Q.; Qu, F.N.; Han, B.; Wang, C.; Li, H.; Mauck, R.L.; Han, L. Micromechanical anisotropy and heterogeneity of the meniscus extracellular matrix. *Acta Biomater.* **2017**, *54*, 356–366. [CrossRef] [PubMed]
44. Anderson, A.T.; Van Houten, E.E.W.; McGarry, M.D.J.; Paulsen, K.D.; Holtrop, J.L.; Sutton, B.P.; Georgiadis, J.G.; Johnson, C.L. Observation of direction-dependent mechanical properties in the human brain with multi-excitation MR elastography. *J. Mech. Behav. Biomed.* **2016**, *59*, 538–546. [CrossRef] [PubMed]
45. Duan, P.F.; Chen, J.J. Nanomechanical and microstructure analysis of extracellular matrix layer of immortalized cell line Y201 from human mesenchymal stem cells. *Surf. Coat. Technol.* **2015**, *284*, 417–421. [CrossRef]
46. Duan, P.F.; Toumpaniari, R.; Partridge, S.; Birch, M.A.; Genever, P.G.; Bull, S.J.; Dalgarno, K.W.; McCaskie, A.W.; Chen, J.J. How cell culture conditions affect the microstructure and nanomechanical properties of extracellular matrix formed by immortalized human mesenchymal stem cells: An experimental and modelling study. *Mater. Sci. Eng. C* **2018**, *89*, 149–159. [CrossRef] [PubMed]
47. Bonilla, M.R.; Stokes, J.R.; Gidley, M.J.; Yakubov, G.E. Interpreting atomic force microscopy nanoindentation of hierarchical biological materials using multi-regime analysis. *Soft Matter* **2015**, *11*, 1281–1292. [CrossRef] [PubMed]
48. Feng, Y.; Okamoto, R.J.; Namani, R.; Genin, G.M.; Bayly, P.V. Measurements of mechanical anisotropy in brain tissue and implications for transversely isotropic material models of white matter. *J. Mech. Behav. Biomed. Mater.* **2013**, *23*, 117–132. [CrossRef] [PubMed]
49. Fischer-Cripps, A. *Nanoindentation*; Springer: New York, NY, USA, 2004.
50. VanLandingham, M.R. Review of instrumented indentation. *J. Res. Natl. Inst. Stand. Technol.* **2003**, *108*, 249–265. [CrossRef] [PubMed]
51. Hoffler, C.E.; Moore, K.E.; Kozloff, K.; Zysset, P.K.; Brown, M.B.; Goldstein, S.A. Heterogeneity of bone lamellar-level elastic moduli. *Bone* **2000**, *26*, 603–609. [CrossRef]
52. Bushby, A.J.; Ferguson, V.L.; Boyde, A. Nanoindentation of bone: Comparison of specimens tested in liquid and embedded in polymethylmethacrylate. *J. Mater. Res.* **2004**, *19*, 249–259. [CrossRef]
53. Wang, S.B.; Liu, H.; Xu, L.X.; Du, X.C.; Zhao, D.; Zhu, B.; Yu, M.; Zhao, H.W. Investigations of Phase Transformation in Monocrystalline Silicon at Low Temperatures via Nanoindentation. *Sci. Rep.* **2017**, *7*, 8682. [CrossRef] [PubMed]
54. Cappella, B.; Dietler, G. Force-distance curves by atomic force microscopy. *Surf. Sci. Rep.* **1999**, *34*, 1–104. [CrossRef]
55. Butt, H.J.; Cappella, B.; Kappl, M. Force measurements with the atomic force microscope: Technique, interpretation and applications. *Surf. Sci. Rep.* **2005**, *59*, 1–152. [CrossRef]
56. Gautier, H.O.B.; Thompson, A.J.; Achouri, S.; Koser, D.E.; Holtzmann, K.; Moeendarbary, E.; Franze, K. Atomic force microscopy-based force measurements on animal cells and tissues. *Method Cell Biol.* **2015**, *125*, 211–235. [CrossRef]
57. Doerner, M.F.; Nix, W.D. A Method for Interpreting the Data form Depth-Sensing Indentation Instruments. *J. Mater. Res.* **1986**, *1*, 601–609. [CrossRef]
58. Oliver, W.C.; Pharr, G.M.J. An Improved Technique for Determining Hardness and Elastic Modulus Using Load and Displacement Sensing Indentation. *J. Mater. Res.* **1992**, *7*, 1564–1583. [CrossRef]

59. Sun, X.D.; Zhao, H.W.; Yu, Y.; Zhang, S.Z.; Ma, Z.C.; Li, N.; Yu, M.; Hou, P.L. Variations of mechanical property of out circumferential lamellae in cortical bone along the radial by nanoindentation. *AIP Adv.* **2016**, *6*, 115116. [CrossRef]

60. Tranchida, D.; Piccarolo, S. On the use of the nanoindentation unloading curve to measure the Young's modulus of polymers on a nanometer scale. *Macromol. Rapid Commun.* **2005**, *26*, 1800–1804. [CrossRef]

61. Cheng, Y.T.; Cheng, C.M. Relationships between initial unloading slope, contact depth, and mechanical properties for conical indentation in linear viscoelastic solids. *J. Mater. Res.* **2005**, *20*, 1046–1053. [CrossRef]

62. Briscoe, B.J.; Fiori, L.; Pelillo, E. Nano-indentation of polymeric surfaces. *J. Phys. D Appl. Phys.* **1998**, *31*, 2395–2405. [CrossRef]

63. Feng, G.; Ngan, A.H.W. Effects of creep and thermal drift on modulus measurement using depth-sensing indentation. *J. Mater. Res.* **2002**, *17*, 660–668. [CrossRef]

64. Tang, B.; Ngan, A.H.W. Accurate measurement of tip-sample contact size during nanoindentation of viscoelastic materials. *J. Mater. Res.* **2003**, *18*, 1141–1148. [CrossRef]

65. Ngan, A.H.W.; Wang, H.T.; Tang, B.; Sze, K.Y. Correcting power-law viscoelastic effects in elastic modulus measurement using depth-sensing indentation. *Int. J. Solids Struct.* **2005**, *42*, 1831–1846. [CrossRef]

66. Shih, C.W.; Yang, M.; Li, J.C.M. Effect of tip radius on nanoindentation. *J. Mater. Res.* **1991**, *6*, 2623–2628. [CrossRef]

67. Chen, W.M.; Li, M.; Zhang, T.; Cheng, Y.T.; Cheng, C.M. Influence of indenter tip roundness on hardness behavior in nanoindentation. *Mater. Sci. Eng. A Struct.* **2007**, *445*, 323–327. [CrossRef]

68. Bolshakov, A.; Pharr, G.M. Influences of pileup on the measurement of mechanical properties by load and depth sensing indentation techniques. *J. Mater. Res.* **1998**, *13*, 1049–1058. [CrossRef]

69. McElhaney, K.W.; Vlassak, J.J.; Nix, W.D. Determination of indenter tip geometry and indentation contact area for depth-sensing indentation experiments. *J. Mater. Res.* **1998**, *13*, 1300–1306. [CrossRef]

70. Calabri, L.; Pugno, N.; Menozzi, C.; Valeri, S. AFM nanoindentation: Tip shape and tip radius of curvature effect on the hardness measurement. *J. Phys. Condens. Matter* **2008**, *20*, 474208. [CrossRef]

71. Chen, J.; Bull, S.J. Relation between the ratio of elastic work to the total work of indentation and the ratio of hardness to Young's modulus for a perfect conical tip. *J. Mater. Res.* **2009**, *24*, 590–598. [CrossRef]

72. Johnson, K. *Contact Mechanics*; Cambridge University Press: Cambridge, UK, 1987.

73. Sneddon, I.N. The relation between load and penetration in the axisymmetric Boussinesq problem for a punch of arbitrary profile. *Int. J. Eng. Sci.* **1965**, *3*, 47–57. [CrossRef]

74. Zhang, M.G.; Chen, J.J.; Feng, X.Q.; Cao, Y.P. On the Applicability of Sneddon's Solution for Interpreting the Indentation of Nonlinear Elastic Biopolymers. *J. Appl. Mech.* **2014**, *81*, 091011. [CrossRef]

75. Li, Y.; Deng, J.X.; Zhou, J.; Li, X.E. Elastic and viscoelastic mechanical properties of brain tissues on the implanting trajectory of sub-thalamic nucleus stimulation. *J. Mater. Sci. Mater. Med.* **2016**, *27*, 163. [CrossRef] [PubMed]

76. Hayes, W.C.; Keer, L.M.; Herrmann, G.; Mockros, L.F. A mathematical analysis for indentation tests of articular cartilage. *J. Biomech.* **1972**, *5*, 541–551. [CrossRef]

77. Finan, J.D.; Fox, P.M.; Morrison, B. Non-ideal effects in indentation testing of soft tissues. *Biomech. Model. Mechanobiol.* **2014**, *13*, 573–584. [CrossRef] [PubMed]

78. King, R.B. Elastic analysis of some punch problems for a layered medium. *Int. J. Solids Struct.* **1987**, *23*, 1657–1664. [CrossRef]

79. Fabrikant, V. Flat punch of arbitrary shape on an elastic half-space. *Int. J. Eng. Sci.* **1986**, *24*, 1731–1740. [CrossRef]

80. Chen, J.J.; Lu, G.X. Finite element modelling of nanoindentation based methods for mechanical properties of cells. *J. Biomech.* **2012**, *45*, 2810–2816. [CrossRef]

81. Darling, E.M.; Zauscher, S.; Block, J.A.; Guilak, F. A thin-layer model for viscoelastic, stress-relaxation testing of cells using atomic force microscopy: Do cell properties reflect metastatic potential? *Biophys. J.* **2007**, *92*, 1784–1791. [CrossRef]

82. Chen, J.J. Nanobiomechanics of living cells: A review. *Interface Focus* **2014**, *4*, 20130055. [CrossRef]

83. Lee, E.H.; Radok, J.R.M. The Contact Problem for Viscoelastic Bodies. *J. Appl. Mech.* **1960**, *33*, 395–401. [CrossRef]

84. Qiu, S.H.; Zhao, X.F.; Chen, J.Y.; Zeng, J.F.; Chen, S.Q.; Chen, L.; Meng, Y.; Liu, B.; Shan, H.; Gao, M.Y.; et al. Characterizing viscoelastic properties of breast cancer tissue in a mouse model using indentation. *J. Biomech.* **2018**, *69*, 81–89. [CrossRef] [PubMed]

85. Mattice, J.M.; Lau, A.G.; Oyen, M.L.; Kent, R.W. Spherical indentation load-relaxation of soft biological tissues. *J. Mater. Res.* **2006**, *21*, 2003–2010. [CrossRef]

86. Kilpatrick, J.I.; Revenko, I.; Rodriguez, B.J. Nanomechanics of Cells and Biomaterials Studied by Atomic Force Microscopy. *Adv. Healthc. Mater.* **2015**, *4*, 2456–2474. [CrossRef] [PubMed]

87. Lin, D.C.; Horkay, F. Nanomechanics of polymer gels and biological tissues: A critical review of analytical approaches in the Hertzian regime and beyond. *Soft Matter* **2008**, *4*, 669–682. [CrossRef]

88. Budday, S.; Nay, R.; de Rooij, R.; Steinmann, P.; Wyrobek, T.; Ovaert, T.C.; Kuhl, E. Mechanical properties of gray and white matter brain tissue by indentation. *J. Mech. Behav. Biomed.* **2015**, *46*, 318–330. [CrossRef] [PubMed]

89. Berdyyeva, T.K.; Woodworth, C.D.; Sokolov, I. Human epithelial cells increase their rigidity with ageing in vitro: Direct measurements. *Phys. Med. Biol.* **2005**, *50*, 81–92. [CrossRef] [PubMed]

90. Oyen, M.L. Nanoindentation of Biological and Biomimetic Materials. *Exp. Tech.* **2013**, *37*, 73–87. [CrossRef]

91. Chudoba, T.; Schwaller, P.; Rabe, R.; Breguet, J.M.; Michler, J. Comparison of nanoindentation results obtained with Berkovich and cube-corner indenters. *Philos. Mag.* **2006**, *86*, 5265–5283. [CrossRef]

92. Kruzic, J.J.; Kim, D.K.; Koester, K.J.; Ritchie, R.O. Indentation techniques for evaluating the fracture toughness of biomaterials and hard tissues. *J. Mech. Behav. Biomed.* **2009**, *2*, 384–395. [CrossRef]

93. Duan, P.F.; Bull, S.; Chen, J.J. Modeling the nanomechanical responses of biopolymer composites during the nanoindentation. *Thin Solid Films* **2015**, *596*, 277–281. [CrossRef]

94. Chen, J.J. On the determination of coating toughness during nanoindentation. *Surf. Coat. Technol.* **2012**, *206*, 3064–3068. [CrossRef]

95. Ozkan, A.D.; Topal, A.E.; Dikecoglu, F.B.; Guler, M.O.; Dana, A.; Tekinay, A.B. Probe microscopy methods and applications in imaging of biological materials. *Semin. Cell Dev. Biol.* **2018**, *73*, 153–164. [CrossRef] [PubMed]

96. Kaufman, J.D.; Klapperich, C.M. Surface detection errors cause overestimation of the modulus in nanoindentation on soft materials. *J. Mech. Behav. Biomed. Mater.* **2009**, *2*, 312–317. [CrossRef] [PubMed]

97. Dimitriadis, E.K.; Horkay, F.; Maresca, J.; Kachar, B.; Chadwick, R.S. Determination of elastic moduli of thin layers of soft material using the atomic force microscope. *Biophys. J.* **2002**, *82*, 2798. [CrossRef]

98. Nicolle, S.; Lounis, M.; Willinger, R. Shear Properties of Brain Tissue over a Frequency Range Relevant for Automotive Impact Situations: New Experimental Results. *Stapp. Car Crash J.* **2004**, *48*, 239–258. [PubMed]

99. Elkin, B.S.; Ilankova, A.; Morrison, B. Dynamic, regional mechanical properties of the porcine brain: Indentation in the coronal plane. *J. Biomech. Eng.* **2011**, *133*, 071009. [CrossRef] [PubMed]

100. Elkin, B.S.; Ilankovan, A.; Morrison, B., 3rd. Age-dependent regional mechanical properties of the rat hippocampus and cortex. *J. Biomech. Eng.* **2010**, *132*, 011010. [CrossRef] [PubMed]

101. Budday, S.; Sommer, G.; Birkl, C.; Langkammer, C.; Haybaeck, J.; Kohnert, J.; Bauer, M.; Paulsen, F.; Steinmann, P.; Kuhl, E.; et al. Mechanical characterization of human brain tissue. *Acta Biomater.* **2017**, *48*, 319–340. [CrossRef] [PubMed]

102. Leipzig, N.D.; Athanasiou, K.A. Unconfined creep compression of chondrocytes. *J. Biomech.* **2005**, *38*, 77–85. [CrossRef] [PubMed]

103. Darling, E.M.; Zauscher, S.; Guilak, F. Viscoelastic properties of zonal articular chondrocytes measured by atomic force microscopy. *Osteoarthr. Cartilage* **2006**, *14*, 571–579. [CrossRef]

104. Gupta, S.; Carrillo, F.; Li, C.; Pruitt, L.; Puttlitz, C. Adhesive forces significantly affect elastic modulus determination of soft polymeric materials in nanoindentation. *Mater. Lett.* **2007**, *61*, 448–451. [CrossRef]

105. Crick, S.L.; Yin, F.C. Assessing micromechanical properties of cells with atomic force microscopy: Importance of the contact point. *Biomech. Model. Mechanobiol.* **2007**, *6*, 199–210. [CrossRef] [PubMed]

106. Carrillo, F.; Gupta, S.; Balooch, M.; Marshall, S.J.; Marshall, G.W.; Pruitt, L.; Puttlitz, C.M. Nanoindentation of polydimethylsiloxane elastomers: Effect of crosslinking, work of adhesion, and fluid environment on elastic modulus. *J. Mater. Res.* **2005**, *20*, 2820–2830. [CrossRef]

107. Johnson, K.L.; Kendall, K.; Roberts, A.D. Surface Energy and the Contact of Elastic Solids. *Proc. R. Soc. Lond. A* **1971**, *324*, 301–313. [CrossRef]

108. Derjaguin, B.V.; Muller, V.M.; Toporov, Y.P. Effect of contact deformations on the adhesion of particles. *J. Colloid Interface Sci.* **1975**, *53*, 314–326. [CrossRef]

109. Tabor, D. Surface forces and surface interactions. *J. Colloid Interface Sci.* **1977**, *58*, 2–13. [CrossRef]

110. Maugis, D. Adhesion of spheres: The JKR-DMT transition using a dugdale model. *J. Colloid Interface Sci.* **1992**, *150*, 243–269. [CrossRef]

111. Dugdale, D.S. Yielding of steel sheets containing slits. *J. Mech. Phys. Solids* **1960**, *8*, 100–104. [CrossRef]

112. Lin, D.C.; Dimitriadis, E.K.; Horkay, F. Robust strategies for automated AFM force curve analysis—I. Non-adhesive indentation of soft, inhomogeneous materials. *J. Biomech. Eng.* **2007**, *129*, 430–440. [CrossRef]

113. Jaasma, M.J.; Jackson, W.M.; Keaveny, T.M. Measurement and characterization of whole-cell mechanical behavior. *Ann. Biomed. Eng.* **2006**, *34*, 748–758. [CrossRef]

114. Rudoy, D.; Yuen, S.G.; Howe, R.D.; Wolfe, P.J. Bayesian change-point analysis for atomic force microscopy and soft material indentation. *J. R. Stat. Soc.* **2010**, *59*, 573–593. [CrossRef]

115. E, A.H.; Heinz, W.F.; Antonik, M.D.; D'Costa, N.P.; Nageswaran, S.; Schoenenberger, C.A.; Hoh, J.H. Relative microelastic mapping of living cells by atomic force microscopy. *Biophys. J.* **1998**, *74*, 1564–1578. [CrossRef]

116. Carl, P.; Schillers, H. Elasticity measurement of living cells with an atomic force microscope: Data acquisition and processing. *Pflugers Arch.* **2008**, *457*, 551–559. [CrossRef] [PubMed]

117. Jeng, Y.R.; Mao, C.P.; Wu, K.T. Instrumented Indentation Investigation on the Viscoelastic Properties of Porcine Cartilage. *J. Bionic Eng.* **2013**, *10*, 522–531. [CrossRef]

118. Yuan, Y.; Verma, R. Measuring microelastic properties of stratum corneum. *Colloids Surf. B Biointerfaces* **2006**, *48*, 6–12. [CrossRef] [PubMed]

119. Strange, D.G.T.; Fletcher, T.L.; Tonsomboon, K.; Brawn, H. Separating poroviscoelastic deformation mechanisms in hydrogels. *Appl. Phys. Lett.* **2013**, *102*, 16–72. [CrossRef]

120. Galli, M.; Comley, K.S.C.; Shean, T.A.V.; Oyen, M.L. Viscoelastic and poroelastic mechanical characterization of hydrated gels. *J. Mater. Res.* **2009**, *24*, 973–979. [CrossRef]

121. Hu, Y.; Zhao, X.; Vlassak, J.J.; Suo, Z. Using indentation to characterize the poroelasticity of gels. *Appl. Phys. Lett.* **2010**, *96*, 121904. [CrossRef]

122. Gupta, S.; Lin, J.; Ashby, P.; Pruitt, L. A fiber reinforced poroelastic model of nanoindentation of porcine costal cartilage: A combined experimental and finite element approach. *J. Mech. Behav. Biomed.* **2009**, *2*, 326–338. [CrossRef]

123. Balooch, M.; Wu-Magidi, I.C.; Balazs, A.; Lundkvist, A.S.; Marshall, S.J.; Marshall, G.W.; Siekhaus, W.J.; Kinney, J.H. Viscoelastic properties of demineralized human dentin measured in water with atomic force microscope (AFM)-based indentation. *J. Biomed. Mater. Res.* **1998**, *40*, 539–544. [CrossRef]

124. Grant, C.A.; Brockwell, D.J.; Radford, S.E.; Thomson, N.H. Effects of hydration on the mechanical response of individual collagen fibrils. *Appl. Phys. Lett.* **2008**, *92*, 233902. [CrossRef]

125. Bembey, A.K.; Bushby, A.J.; Boyde, A.; Ferguson, V.L.; Oyen, M.L. Hydration effects on the micro-mechanical properties of bone. *J. Mater. Res.* **2006**, *21*, 1962–1968. [CrossRef]

126. Binnig, G.; Quate, C.F.; Gerber, C. Atomic force microscope. *Phys. Rev. Lett.* **1986**, *56*, 930–933. [CrossRef]

127. Dufrene, Y.F. Using nanotechniques to explore microbial surfaces. *Nat. Rev. Microbiol.* **2004**, *2*, 451–460. [CrossRef] [PubMed]

128. Hinterdorfer, P.; Dufrene, Y.F. Detection and localization of single molecular recognition events using atomic force microscopy. *Nat. Methods* **2006**, *3*, 347–355. [CrossRef] [PubMed]

129. Garcia, R.; Perez, R. Dynamic atomic force microscopy methods. *Surf. Sci. Rep.* **2002**, *47*, 197–301. [CrossRef]

130. Jalili, N.; Laxminarayana, K. A review of atomic force microscopy imaging systems: Application to molecular metrology and biological sciences. *Mechatronics* **2004**, *14*, 907–945. [CrossRef]

131. Guzman, C.; Jeney, S.; Kreplak, L.; Kasas, S.; Kulik, A.J.; Aebi, U.; Forro, L. Exploring the mechanical properties of single vimentin intermediate filaments by atomic force microscopy. *J. Mol. Biol.* **2006**, *360*, 623–630. [CrossRef] [PubMed]

132. Kim, J.S.; Park, Y.S.; Nam, H.G.; Park, J.W. Imaging a specific mRNA in pollen with atomic force microscopy. *RSC Adv.* **2015**, *5*, 18858–18865. [CrossRef]

133. Schon, P. Atomic force microscopy of RNA: State of the art and recent advancements. *Semin. Cell Dev. Biol.* **2018**, *73*, 209–219. [CrossRef]

134. Kasas, S.; Dietler, G. DNA-protein interactions explored by atomic force microscopy. *Semin. Cell Dev. Biol.* **2018**, *73*, 231–239. [CrossRef] [PubMed]

135. Dumitru, A.C.; Herruzo, E.T.; Rausell, E.; Cena, V.; Garcia, R. Unbinding forces and energies between a siRNA molecule and a dendrimer measured by force spectroscopy. *Nanoscale* **2015**, *7*, 20267–20276. [CrossRef] [PubMed]

136. Willner, I.; Willner, B. Biomolecule-Based Nanomaterials and Nanostructures. *Nano Lett.* **2010**, *10*, 3805–3815. [CrossRef] [PubMed]

Micromachines **2018**, *9*, 654

137. Shieh, A.C.; Athanasiou, K.A. Principles of cell mechanics for cartilage tissue engineering. *Ann. Biomed. Eng.* **2003**, *31*, 1–11. [CrossRef] [PubMed]
138. Zemla, J.; Danilkiewicz, J.; Orzechowska, B.; Pabijan, J.; Seweryn, S.; Lekka, M. Atomic force microscopy as a tool for assessing the cellular elasticity and adhesiveness to identify cancer cells and tissues. *Semin. Cell Dev. Biol.* **2018**, *73*, 115–124. [CrossRef] [PubMed]
139. Guo, Q.Q.; Xia, Y.; Sandig, M.; Yang, J. Characterization of cell elasticity correlated with cell morphology by atomic force microscope. *J. Biomech.* **2012**, *45*, 304–309. [CrossRef] [PubMed]
140. Smith, J.R.; Tsibouklis, J.; Nevell, T.G.; Breakspear, S. AFM friction and adhesion mapping of the substructures of human hair cuticles. *Appl. Surf. Sci.* **2013**, *285*, 638–644. [CrossRef]
141. Benoit, M.; Gabriel, D.; Gerisch, G.; Gaub, H.E. Discrete interactions in cell adhesion measured by single-molecule force spectroscopy. *Nat. Cell Biol.* **2000**, *2*, 313–317. [CrossRef]
142. Rianna, C.; Kumar, P.; Radmacher, M. The role of the microenvironment in the biophysics of cancer. *Semin. Cell Dev. Biol.* **2018**, *73*, 107–114. [CrossRef]
143. Alcaraz, J.; Otero, J.; Jorba, I.; Navajas, D. Bidirectional mechanobiology between cells and their local extracellular matrix probed by atomic force microscopy. *Semin. Cell Dev. Biol.* **2018**, *73*, 71–81. [CrossRef]
144. Shi, Y.; Cai, M.J.; Zhou, L.L.; Wang, H.D. The structure and function of cell membranes studied by atomic force microscopy. *Semin. Cell Dev. Biol.* **2018**, *73*, 31–44. [CrossRef]
145. Bitler, A.; Dover, R.S.; Shai, Y. Fractal properties of cell surface structures: A view from AFM. *Semin. Cell Dev. Biol.* **2018**, *73*, 64–70. [CrossRef] [PubMed]
146. Borin, D.; Pecorari, I.; Pena, B.; Sbaizero, O. Novel insights into cardiomyocytes provided by atomic force microscopy. *Semin. Cell Dev. Biol.* **2018**, *73*, 4–12. [CrossRef] [PubMed]
147. Braet, F.; Taatjes, D.J.; Wisse, E. Probing the unseen structure and function of liver cells through atomic force microscopy. *Semin. Cell Dev. Biol.* **2018**, *73*, 13–30. [CrossRef]
148. Jin, H.; Liang, Q.; Chen, T.S.; Wang, X.P. Resveratrol Protects Chondrocytes from Apoptosis via Altering the Ultrastructural and Biomechanical Properties: An AFM Study. *PLoS ONE* **2014**, *9*, e91611. [CrossRef] [PubMed]
149. Elkin, B.S.; Azeloglu, E.U.; Costa, K.D.; Morrison, B. Mechanical heterogeneity of the rat hippocampus measured by atomic force microscope indentation. *J. Neurotraum.* **2007**, *24*, 812–822. [CrossRef] [PubMed]
150. Braunsmann, C.; Seifert, J.; Rheinlaender, J.; Schaffer, T.E. High-speed force mapping on living cells with a small cantilever atomic force microscope. *Rev. Sci. Instrum.* **2014**, *85*, 073703. [CrossRef] [PubMed]
151. Maloney, J.M.; Nikova, D.; Lautenschlager, F.; Clarke, E.; Langer, R.; Guck, J.; Van Vliet, K.J. Mesenchymal Stem Cell Mechanics from the Attached to the Suspended State. *Biophys. J.* **2010**, *99*, 2479–2487. [CrossRef] [PubMed]
152. O'Connor, W.T.; Smyth, A.; Gilchrist, M.D. Animal models of traumatic brain injury: A critical evaluation. *Pharmacol. Ther.* **2011**, *130*, 106–113. [CrossRef]
153. MacManus, D.B.; Murphy, J.G.; Gilchrist, M.D. Mechanical characterisation of brain tissue up to 35% strain at 1, 10, and 100/s using a custom-built micro-indentation apparatus. *J. Mech. Behav. Biomed.* **2018**, *87*, 256–266. [CrossRef]
154. Wahlquist, J.A.; DelRio, F.W.; Randolph, M.A.; Aziz, A.H.; Heveran, C.M.; Bryant, S.J.; Neu, C.P.; Ferguson, V.L. Indentation mapping revealed poroelastic, but not viscoelastic, properties spanning native zonal articular cartilage. *Acta Biomater.* **2017**, *64*, 41–49. [CrossRef] [PubMed]
155. Liu, D.; Li, G.Y.; Su, C.; Zheng, Y.; Jiang, Y.X.; Qian, L.X.; Cao, Y.P. Effect of ligation on the viscoelastic properties of liver tissues. *J. Biomech.* **2018**, *76*, 235–240. [CrossRef] [PubMed]
156. Liang, K.N.; Xiao, S.M.; Shi, W.Y.; Li, J.S.; Yang, X.; Gao, Y.; Gou, Y.P.; Hao, L.Y.; He, L.B.; Cheng, L.; et al. 8DSS-promoted remineralization of demineralized dentin in vitro. *J. Mater. Chem. B* **2015**, *3*, 6763–6772. [CrossRef]
157. Boyer, G.; Laquieze, L.; Le Bot, A.; Laquieze, S.; Zahouani, H. Dynamic indentation on human skin in vivo: Ageing effects. *Skin Res. Technol.* **2009**, *15*, 55–67. [CrossRef] [PubMed]
158. Chatelin, S.; Constantinesco, A.; Willinger, R. Fifty years of brain tissue mechanical testing: From in vitro to in vivo investigations. *Biorheology* **2010**, *47*, 255–276. [CrossRef] [PubMed]
159. Donnelly, E.; Baker, S.P.; Boskey, A.L.; van der Meulen, M.C.H. Effects of surface roughness and maximum load on the mechanical properties of cancellous bone measured by nanoindentation. *J. Biomed. Mater. Res. A* **2006**, *77*, 426–435. [CrossRef] [PubMed]

160. Chen, J.J.; Birch, M.A.; Bull, S.J. Nanomechanical characterization of tissue engineered bone grown on titanium alloy in vitro. *J. Mater. Sci. Mater. Med.* **2010**, *21*, 277–282. [CrossRef] [PubMed]

161. Goriely, A.; Geers, M.G.D.; Holzapfel, G.A.; Jayamohan, J.; Jerusalem, A.; Sivaloganathan, S.; Squier, W.; van Dommelen, J.A.W.; Waters, S.; Kuhl, E. Mechanics of the brain: Perspectives, challenges, and opportunities. *Biomech. Model. Mechanobiol.* **2015**, *14*, 931–965. [CrossRef]

162. Sokolov, I.; Dokukin, M.E.; Guz, N.V. Method for quantitative measurements of the elastic modulus of biological cells in AFM indentation experiments. *Methods* **2013**, *60*, 202–213. [CrossRef]

163. Weickenmeier, J.; Kurt, M.; Ozkaya, E.; de Rooij, R.; Ovaert, T.C.; Ehman, R.L.; Pauly, K.B.; Kuhl, E. Brain stiffens post mortem. *J. Mech. Behav. Biomed.* **2018**, *84*, 88–98. [CrossRef]

164. Qian, L.; Sun, Y.F.; Tian, J.Y.; Ren, Z.; Zhao, H.W. Indentation response of porcine brain under electric fields. *Soft Matter* **2018**, under review.

165. Azeloglu, E.U.; Costa, K.D. Atomic force microscopy in mechanobiology: Measuring microelastic heterogeneity of living cells. *Methods Mol. Biol.* **2011**, *736*, 303–329. [CrossRef] [PubMed]

166. Jacobs, C.; Huang, H.; Kwon, R.Y. *Introduction to Cell Mechanics and Mechanobiology*; Garland Science: New York, NY, USA, 2012.

167. Oudry, J.; Lynch, T.; Vappou, J.; Sandrin, L.; Miette, V. Comparison of four different techniques to evaluate the elastic properties of phantom in elastography: Is there a gold standard? *Phys. Med. Biol.* **2014**, *59*, 5775–5793. [CrossRef] [PubMed]

MDPI

St. Alban-Anlage 66

4052 Basel

Switzerland

Tel. +41 61 683 77 34

Fax +41 61 302 89 18

www.mdpi.com

Micromachines Editorial Office

E-mail: micromachines@mdpi.com

www.mdpi.com/journal/micromachines

www.ingramcontent.com/pod-product-compliance
Lightning Source LLC
Chambersburg PA
CBHW041217220326
41597CB00033BA/5994